PRECISION SHOOTING WITH THE
M1 GARAND

by
Roy Baumgardner

Published by:
Precision Shooting, Inc.
222 McKee Street
Manchester, CT 06040

First Edition: September, 1999
Second Edition: December, 2000
Third Edition: February, 2003

(Many of these articles appeared previously in either *Precision Shooting* or *Tactical Shooter* magazines).

About the cover. . .

Joseph Jacob Foss has lived more and accomplished more in 84 years than most people could hope for in two lifetimes. Born in rural South Dakota in 1915, he graduated from the University of SD in 1940, by which time he was already a licensed pilot. He entered naval flight training later that year, receiving a Marine Corps commission in March 1941. By October 1942 he was a Captain and Executive Officer of Marine Fighter Squadron 121, flying Grumman Wildcats at Guadalcanal in the Solomon Islands. Upon leaving combat in January 1943 he was America's leading fighter ace with 26 credited aerial victories, a feat that earned him the Medal of Honor.

Joe formed his own squadron, VMF-115, and returned to the Pacific in 1944, but malaria forced a premature return home. Owing to a bureaucratic snafu, he missed the chance for a regular commission but was scooped up by the Air Force Reserve. He was cofounder of the South Dakota Air National Guard and attained Brigadier General rank in 1954. Meanwhile, he was elected to the State House, two terms as Governor, and led his own aerobatic team.

From 1959–66 Joe was first Commissioner of the American Football League, and he hosted two TV programs. He was NRA President 1988–90 and has been a Board member since then. He has co-authored two autobiographies and an anthology of fighter aviation, and remains a much sought-after lecturer between his frequent hunting trips and work on behalf of crippled children and Campus Crusade.

Photo and text courtesy of Barrett Tillman

Dedication

The book is dedicated, with great respect, to all the men who ever carried an M1 Garand Rifle . . . whether it was on the battlefields of war, or the target ranges of peace.

Roy Baumgardner

"In my opinion, the M1 Rifle is the greatest battle implement ever devised."

Lt. Gen. George S. Patton, Jr.
January 26, 1945

Table of Contents

Introduction

A high percentage of the Highpower shooters throughout the length and width of the land appear to be in love with the venerable M1 Garand these days. Why, in a sport where cutting-edge new developments are immediately in great demand... and hang the expense... why the retrogression to a rifle designed in the mid-1930s, and which has since been replaced in the U.S. military by the M14, and that in turn has been replaced by the M16?

Probably there are a number of factors involved in the uncanny popularity of the M1 Garand today. First, of course, is the fact that they are widely available, both on the general firearms-for-sale scene, and also via the CMP at a discount price level. Second would probably be the fact that in today's firearms climate, the M1 Garand simply does not look either particularly alarming or lethal... whereas some of the military weapons of the day, of the "assault rifle" genre, tend to cause widespread alarm among the citizenry even at a distance, and hysteria when up close. Millions of our military became familiar with the Garand in the days of World War II, the Korean War, and for many years after in both the active armed forces and the National Guard... thus today there are thousands of individuals who can pick up a Garand with some degree of familiarity, as opposed to the empty feeling of facing a totally strange mechanical device. Since memories have a tendency to grow better with age... the reunion between long-ago armed forces types, and an M1 Garand, often takes on a reunion-with-an-old-friend scenario.

To quote from the June, 1999 issue of *Shooting Sports USA,* where that publication talks about the new M1 Garand "class"... "In fact, this new class is not a class at all, but a category for as-issued highpower rifles. The category was designed as a complement to the JC Garand match at Camp Perry. Its intent is to provide new shooters with limited equipment, the opportunity to experience competition before deciding whether or not to make a more moneyed investment in tournaments and gear. Guidelines created by the NRA will be provided to those wishing to be match sponsors of this as-issued M1 Garand category. The CMP is sponsoring this event at Camp Perry and it will be run by the NRA."

Today the M1 Garand still has a lot of admirers and enthusiasts as we head into a new millennium, with a 65-year-old design rifle. Still, pretty is as pretty does, and there is a lot to be said, even today, for the combination of the United States' great battle rifle of the 20th century, the ubiquitous .30-06 cartridge (about which the late Townsend Whelen remarked, "The .30-06 is never a mistake"), and the memories of days-gone-by, of a simpler, less frenetic world... remind us of the motto of our friends at Imagine Art Studio, in Virginia... "Because some things are worth going back for."

If you had become acquainted with the Garand earlier in your life, or if you missed making that acquaintance... it's no matter. In this case at least... opportunity is knocking twice for you!

Dave Brennan, Editor
Precision Shooting magazine

Low Cost Accurizing Tips for the M1

Part 1

The Rifle, The Gas Cylinder

(*Precision Shooting,* September 1993)

The M1 Garand is in a rather unique situation today. A few years back it was the top rifle in war, and .30 cal. service rifle competition. Today, it is introducing a very large number of new shooters to big bore target shooting. I also think it would be fair to say that their number is only exceeded by the number of people who just plain want one, whether for collecting, or just because.

At the DCM Clinic, which is a school put on at Camp Perry each spring by the Ohio Rifle and Pistol Association to teach students the National Match course of fire with service rifles, the M1 is used. It usually draws over 500 students. If you wish to enroll, write Barb Smith, PO Box 11490, Columbus, OH 43211-1490. The cost is very modest, about $35.00 for 2 days for an adult.

Some people go through the red tape and get their M1 through the DCM (Director of Civilian Marksmanship) program in Washington

Using the Garand is a very low cost way to get into Highpower competition.

DC. This is a very low cost way to get an M1. For more information on this, write the NRA, 11250 Waples Mill Road, Fairfax, VA 22030-9400. Still others go to dealers and gun shows until they find what they are looking for.

If you get an M1 through the DCM, you can get into service rifle competition for much less than the cost of an AR-15. Consider these rough cost figures. A match grade M1 can run around $900 to $1400, depending on new, used, and who built it; a good AR-15 $600 to $1000; and a good M1 Garand about $450 at a gun show to $165 through the DCM

My wife and I each have DCM rifles. They sent me one made by Harrington and Richardson, and rebuilt at Springfield (Govt. arsenal). They sent her one made at Springfield. It was brand new, a real dazzler in mint condition, with cleaning tools and sling, and the whole nine yards. I thought... I... was... gonna... die. For about the first 200 rounds of GI 150 gr. ammo, the best I could get out of them was 5" groups. Then they settled down to about 3 MOA. I then began learning from some of the armorers and competitors, little things to do to make them shoot tighter groups. It is surprising, how little it took. They still have their original GI barrels, and now shoot around 1½" to 2" MOA groups. This is nothing to shout about, but it should be good enough for the NRA marksman classification, and possibly up into the sharpshooter class, depending on where you shoot, and who you shoot against. The H.R. took a 2nd place medal at 600 yards in the Army Cup match at Perry Nationals in the marksman class a few years back.

The Garand is a good rifle. If you work it down to 1½ to 2 MOA it can hold the 10 ring of the 200 yard "SR" target used in NRA and government matches. Some will shoot under 2 MOA What I will do here is pass along the little tips, and bits of accurizing information that all of the armorers, shooters, and gunsmiths passed on to me. These people are the kindest, finest people in the world. Without them we would be lost. These tips will arguably be in the order in which they will do the most good. They are low in cost if you do them yourself. But they will not take the place of sending your rifle to a top gunsmith and having him go all over it and fit a top of the line barrel. It is important to understand this.

First, is the front sight. Since the Garand was originally designed to be used with corrosive ammunition, the gas cylinder was made a loose

The M1's gas cylinder with lock and nut. Always wiggle it to check for a loose fit on the barrel.

fit on the barrel so that a soldier could get it off to clean it no matter what. It also had to be interchangeable with all the others out there. Since the rifle would not repeat if the gas system was full of rust, and reliability was of more importance than accuracy, the splines on the barrel are a very generous fit on the gas cylinder. Unfortunately Garand put the front sight on top of the gas cylinder. So if you grab most M1s by the front sight, you can wiggle them quite a bit. This of course, is not conducive to good target accuracy.

There are a couple of ways to cure this. Most of the people I talked to just remove the gas cylinder, turn a ball peen hammer upside down, set the round peening end in the slots in the barrel, and give it a couple of taps with another hammer. They do this in two or three places on each of the three splines where the gas cylinder will locate. They then tap the gas cylinder back into place on the rifle. This does indeed work, and makes for a very snug sight.

Still, I have talked to others who state that you should have a soft steel, copper or brass (take your pick), snug fitting spud in the bore before peening to prevent deforming the bore. The first group says that this is not necessary as you are just lightly tapping, to slightly deform the splines, not hammering hard enough to deform the bore, and if you ever saw the way these barrels were handled during manufacturing in wartime, you probably would have trouble getting your breath for awhile.

Still another group feels that the whole problem can be avoided by just fitting the thing with high temp epoxy, that will stand the heat of rapid fire. But don't forget the release agent, and don't get any epoxy into the gas port or bore. I have never tried this, so can't comment on it.

I made a thumb press fitted soft steel spud and pushed it in the bore and then peened the splines lightly, trying the gas cylinder as I went. It worked fine, and the spud had the same amount of push, pull, resistance before as it did after. So the choice is up to you. Any time you get into something like this, it is bound to get controversial. Of course, if you had a good custom barrel fitted, the splines would be tight to begin with.

But then, the argument goes, you have not increased your shooting skills by wearing out the original barrel. I have one friend who has a superb custom barrel on his M1A. He really grits his teeth every time he has to run a string of rapid fire through it. And a double string in the regional course just about kills him, as he is convinced he is wearing out his 600 yard barrel. But he can't hold MOA offhand, so I don't know where the answer lies.

Once the gas cylinder is fitted, there is another little trick that the armorers do to hold it right where they want it (which is centered over the gas port hole in the barrel); and that is to fit different gas cylinder locks, (this is the figure 8 shaped part that screws on to the barrel in front of the gas cylinder) until they set one that will not screw on more than 10° to 30° past the final tight position. They then back it off until everything lines up, and the gas cylinder nut will go in and tighten. In other words, if everything goes together at the six o'clock position, they don't want the lock to tighten past the seven o'clock position with the lock screw removed.

To get this set up, they try fitting different locks until they find one that works. I only had a few thousandths to go on mine to get this, so I just ground it off the back by holding it against the side of a grinding wheel; this worked just fine, and overall, tightened the groups up quite a bit. Since the front sight was no longer shifting from side to side, a lot of shots that went out at three o'clock and nine o'clock that were thought to be caused by wind drift, or bad judgment thereof, were eliminated. There is nothing quite like nonrepeatable lost motion in a

Mr. Garand's rifle. It is surprising how many of them are showing up at ranges around the country.

sight, whether inside an expensive benchrest scope, in the fine threads of an Olympic-type peep sight, or in a minutely loose M1 Garand gas cylinder to drive otherwise perfectly normal target shooters to want to tell waitresses at restaurants what a really lousy day they have had at the range when they go in for supper.

When you stop to think that one click on the windage knob of an M1 rear sight is supposed to move the sight aperture only .008 (eight thousandths of an inch, and often done) to get one minute of angle change, and the average human hair is about .003, you begin to see how you can get big groups, and occasional flyers, from a rifle that is capable of far better accuracy, if the front sight is slightly loose.

It is our government's policy now, and has been for some time, thank God, to slowly empty its warehouses and storage facilities of surplus M1 Garands and to sell them to this nation's trying-hard target shooters. This does not even begin to take into account how many thousands of rifles were sold, or given, to allied countries around the world who are now selling them back to our free market dealers. So it is hard to say how many M1s are out there, but when you stop to think of all the people we supplied from World War II to Vietnam, and that over 4 million were produced, it has got to add up to a lot.

To kind of sum up here, it appears that there are a lot of people out there playing with Mr. Garand's rifle. Some of them get into target shooting and then drop out. Some turn to cast bullet shooting. Some compete with their M1 for awhile, then go on to a match grade AR-15 or M1A, leaving their M1 for practice, or to loan to wife, girlfriend, or buddy in the hope that they will find as much enjoyment with shooting, and the people that they shoot with, as they have spending weekends at the range.

Then, you get the casual shooters, and collectors. These people so outnumber target shooters and serious hunters, that it is hard to believe. You have only to go to any gun show to see how many hundreds of them there are trying to edge past each other in the aisles. Most of those who like the M1, whether for its target shooting, or for its romance, and the part it has played in the changing of our history, have one thing in common. They would like to make it more accurate without spending any money. While this is not 100% possible, it is possible to use a lot of the little tricks passed on to us by armorers, gunsmiths, and just plain tinkerers and experimenters, to make the rifle perform better than it did when it came from the factory. It is hoped that these people will then realize just what the rifle is really capable of, and then put it in the hands of a truly good gunsmith, pay what it is worth, and have a rifle that they will be proud to own forever. Because, the search for accuracy never stops.

Low Cost Accurizing Tips for the M1

Part 2

The Muzzle Crown, the Gas Cylinder Nut and the Sling

(*Precision Shooting,* January 1994)

I have heard many people complain about the red tape, fingerprinting and delay involved in getting a Garand through the DCM program. There are of course perfectly logical reasons for all of this. Take the number of shots fired for record, for instance. If they did not have that requirement the good people who just plain want one, and who far outnumber the .30 caliber Highpower shooters, would leave few left for the target shooters. That is good or bad, depending on your point of view. I, for one, feel that we are very fortunate to even have a program like this. If you ever get involved with some of the other programs out of Dizzy Land, DC you will understand what I mean.

When we left off in part one of this series, we were going over simple low cost modifications that can be made to the rifle's gas cylinder to improve accuracy.

There are some other things that can be done to the gas cylinder also, such as facing off the back of it, and opening up the rear support ring about .010". But if you do this, you should permanently attach the free floated upper handguard to the lower band too. This will help shrink the groups a bit more, but at a price. You will now have an M1 rifle that cannot be grabbed by the front handguard, which is a perfectly normal thing to expect people to do. To do so risks breaking the joint, or bond, between the handguard and the lower band.

So what is suggested is that we leave the gas cylinder and handguards for now, and go on to more simple things based on improving the M1, which will deal with cutting the barrel muzzle crown, how two groups do it slightly different from each other, what is good or bad about each method, and why they chose to do it that way.

Later, we will come back to the single attaching point front handguard and gas cylinder, to discuss their good and not so good points. Basically the idea is good; it just depends on different points of view, and how the rifle will be used. Like many modifications on the M1, it is reversible by buying a couple of replacement parts, or can be done only part way.

Also, be sure to keep the stacking swivel on the gas cylinder tightened to prevent its swinging and hitting the barrel during recoil. It is known that this affects accuracy like a small hammer hitting the muzzle just as the bullet leaves it. Just another small thing.

The other thing to check at this time is the crown at the muzzle of the barrel. If you could ever see the speed at which this counter sink is put on during wartime manufacture by the lowest bidder counting cost in units per hour, you could easily understand how benchrest gunsmiths can turn to strong drink, or restoring sports cars and old furniture.

The crown of Mr. Garand's rifle. You are kidding yourself if you try to get accuracy without having this checked first.

Then consider how many times that muzzle has been banged into a hard surface from the time it left the factory, how many people the rifle may have been issued to, and how it was cared for until now, when you are trying to compete in target matches with it. In spite of all of this, I have seen some that dial indicated less than a 1/2 thousandth.

The best way by far is to pay what it is worth and have a precision gunsmith put the barreled action's muzzle through the 4 jaw chuck on his lathe, indicated off of the lands and grooves, and put a new crown on it if it needs it. I prefer the method of using the tool post to hold the tool, rather than the tailstock, as it can use a positive stop that gives a more precision clean up at the end of the cut. This is more important than once thought.

I have seen armorers at Perry use a "T" handled countersink with a long pilot to touch up service rifles with nicks in the crown. This is

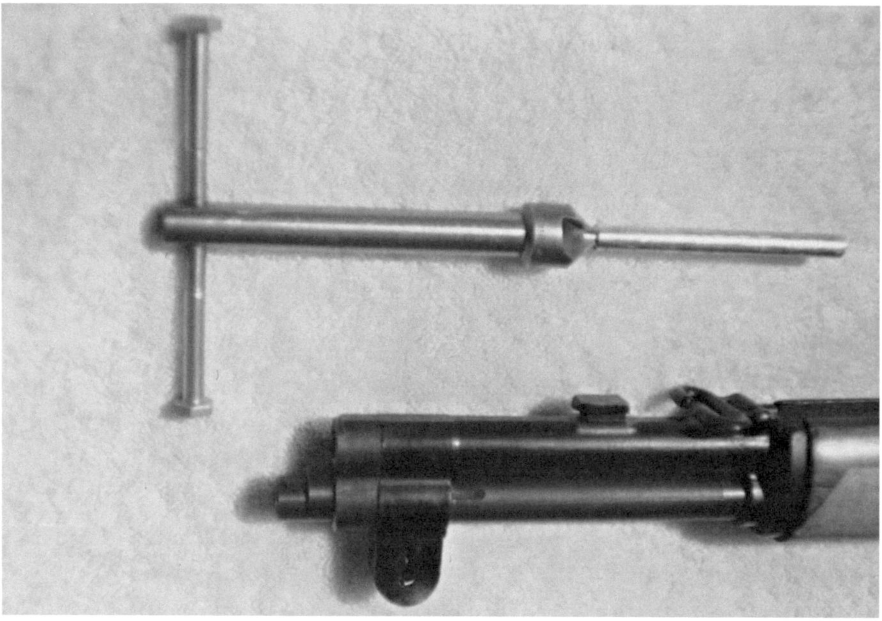

Some use a "T" handled, long pilot countersink to touch up the muzzle crown. It is not as accurate as a positive stop tool, but is better than nothing if needed.

better than nothing, and cheaper than a lathe, especially if no lathe is available; but it does not have a positive stop as a lathe does. You will never really know how good your GI barrel is until everything on it, wood fit, operating rod fit, crown, etc., is performing the way it is supposed to.

One last thing about the gas cylinder. There are two common types of gas cylinder nut. One is solid, and is easily recognized as it has a single screw slot in its face, for a common screwdriver. The other has what appears at first glance to be a Phillips screwdriver slot, but will take an ordinary screwdriver. This one contains a spring loaded valve that is forced open when a grenade launcher is fitted to the rifle so that the gas from the special blank cartridge used to launch the grenade is exhausted out the front and does not go back to hit the op rod. Years ago

M1 gas cylinder nuts. The one with the "X" type slot contains a one-way check valve.

when the M1 was winning everything, it was considered wise to fit the solid one so that there was no valve to fail in the middle of a match — especially if you were well on your way to the top score of the Nationals, and now had to shoot that score over again, if at all possible.

These valves are very trouble free because unless a grenade launcher is fitted, the parts inside never move. The gas from the fired round just pushes against the closed face of the closed valve and it might just as well be solid. About the only thing that might give trouble is the stiff spring that holds the valve closed, so it is not a bad idea to put a few drops of oil in the screwdriver slot to keep it from getting rusty. This also oils the seat of the closed valve.

The next item is the choice of the sling. You can use either the GI sling, the web, or the leather. This is so simple to some that it is almost funny, but it is not funny to the people who for one reason or another have become very very set in their ways about the choice of the sling. If you want a real controversy, suggest to someone who is winning with a web that they switch to a leather one.

Most of the top service rifle shooters who compete with the M14/M1A use the leather sling. Often it is a very thick custom-made one that has large keepers but still meets the service rifle specs and rules.

There are many copies of the Government model sling. Some are very thin to save cost, and many are down-right poor. The Government

Some make the loop for the arm in the back side of the buckle, not the end of the web sling. It is faster to get in and out of.

model is 1½" wide. The copies are available in anywhere from ⅞" to 1" and close metric equivalent to 1¼" Often cheap ones stated to be 1¼" simply are not.

The top shooter carefully decides where on his arm he wants the sling, and exactly how much tension he wants on it, based on his position (sitting, prone, etc.); how much clothing he has on under his shooting coat; and what works best for him. Some shooters do better than others with more or less tension, and what is often overlooked is that some rifles also do the same. Later, when bedding is considered, you will note that M1 Garands are usually bedded so the pressure of the stock ferrule on the lower band falls within 20 to 40 lbs. Since the sling swivel attaches to the stock ferrule, tension here (and whether straight down, or at an angle to one side) can make a difference. Also bear in mind that the lower band is just that, a band around the rifle's barrel. So with all of wood handguards, operating rod, and gas cylinder hanging on the barrel, damping its vibrations, it is a small wonder that these rifles can be made to shoot well at all. This is also one of the reasons that some highpower shooters who have tried both service rifles (called gas guns) and bolt action target rifles (bolt guns) give up completely on service rifles and will not compete with anything but bolt rifles. In spite of all this, service rifles turn in some superb scores.

The double holes for the hooks (frogs) on the Government model leather sling are normally spaced 1¼" apart for sling tension adjustment, and this is considered sufficient by most top shooters, whether using a bolt rifle or a service rifle. However one of the first things that web sling lovers (now there's a catch phrase) will tell you is that adjustment on theirs is infinite; they can set it anywhere. Many draw lines across their sling with a magic marker, or crayon, or ball point pen, to mark where they want the metal keeper to be locked. Also, like the leather sling shooters, they may change this for different positions. One shooter told me he had to change his when he gained too much weight.

I have seen leather slings with extra holes punched in them for closer adjustment, and I have also heard other people say that this weakens the sling. Some like their leather sling soft and pliable, and some like theirs firm and stiff. A top of the line custom leather sling can run about $40.00 and is worth every penny of it if it helps you win. A good Government model can run around $20.00 to $24.00 and a cheap copy about $15.00, whereas a web one is about $3.00 to $5.00. But what it all boils down to is what works for you. The search for accuracy never stops.

Low Cost Accurizing Tips for the M1

Part 3

Epoxy Bedding the Action (And The Different Methods Used)

(*Precision Shooting,* February 1994)

There are so many things that affect the accuracy of the M1, it is hard to point to any one thing, and say, "This is the most important". But one thing is certain, the action cannot be allowed to change position with each shot.

The action to stock fit is pretty loose, the way it comes from the factory. This is mostly so that the parts can interchange with all of the others out in the field, and one of the reasons some rifles tend to shoot better after a few hundred rounds of recoil have pounded them back as far as they will go.

The M1 is still a good rifle to start out with in Highpower. Here, instructor John Wynn adjusts the sling for a student at the DCM clinic at Camp Perry.

Basically, what the gunsmiths are after, is to bed the rifle firmly into its stock, as close to its center line as possible, have plenty of clearance for the operating rod, (filing out for extra clearance where necessary) and have the pressure of the stock ferrule on the lower band fall within 20 to 40 lbs. Some have said that this is somewhat like having pressure at the fore-end tip of a sporting rifle.

The rifle is stripped down to just the barrel and action, only the rear sight is left on. A 1/4" radius is filed, or ground on the bottom rear receiver tangs. This is to help keep them from gouging the bedding when the rifle is later assembled, and disassembled, because then it goes together in a kind of arc.

The trigger housing group is stripped, but the trigger guard, its pin, and the safety are left on. All the holes and clearance cuts in the action and trigger housing are filled with modeling clay, and all metal coated with the release agent of the bedding compound you decide to use. Some gunsmiths mask off the stock with masking tape. One I talked to uses automobile paste wax for an agent, and coats everything he doesn't want the compound to stick to. But, he cautions to be sure and let it dry thoroughly before fitting. Whatever you use, be sure to follow the manufacturer's directions. You make the decision on doing any of this, and if you screw up, like most of us do sooner or later, you are strictly on your own. (Now there is a disclaimer anyone can understand.)

This brings us to the pros and cons of different bedding compounds. Some use aluminum based compounds, some use steel. There has been some talk of the steel rusting against the metal parts in very wet climates if the parts are not well protected, and I have also heard people say not to coat the parts with grease as it allows the action to slip around. Since I was finishing my stock with Birchwood Casey's Tru-Oil, I went with the paste wax and the Micro Sight company's Micro-Bed, feeling that it was harder than the wood of the stock, but easy to cut or file if I screwed up, or changed my mind later. I have not been sorry. By the way, some people also swear by fiberglass.

Exactly how to bed this action is also controversial. Some rout out the wood everywhere it beds against the metal. Some fully bed the floor plate and trigger guard, and rear of the action tangs, but on the top, bed only the rear and two front bottoms of the action so that it sits stable, like a tripod. There is another type of thinking though that is very interesting. It goes like this: Since the wood to metal fit on a GI rifle is already generous, if everything lines up, (and if it doesn't, remove only enough to make it so), then coat everything with epoxy, the top, the inside, and the bottom, and when it sets up, it will all be one big lump, like a sleeved benchrest gun, and this will cancel out all vibrations and movement, yet the rifle can still be readily disassembled if necessary. No routing necessary.

When I heard of this second method, my heart went pitty-pat like it always does when I hear of something new, and I figured that if it didn't

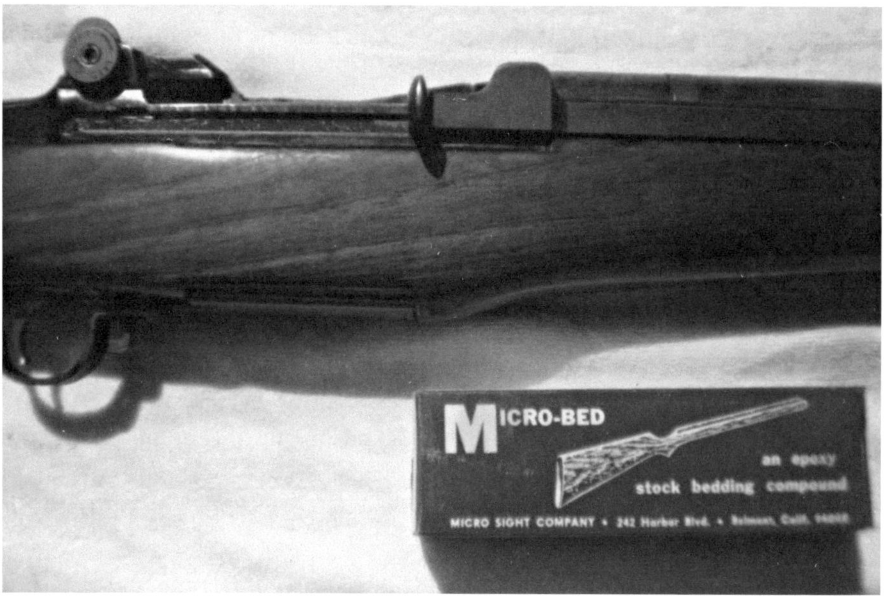

Micro-bed works well because if you screw up, unlike hard glass, it is easy to correct.

work it would not be hard to remove all the Micro-Bed and do it the other way. It worked out fine.

To get the proper tension of the stock ferrule (tip) on the lower band, gunsmiths who do a lot of stocks use small fixtures or bushings that align the barrel in the center of the stock and the correct height above the ferrule. Not having these, another way is to put the barreled action in the stock and see how the barrel lines up above the ferrule. The ferrule is the metal tip at the fore-end of the stock that the front sling swivel attaches to. (This is not to be confused with the stacking swivel which attaches to the gas cylinder.)

The barrel should center above the ferrule, and the distance between the bottom of the barrel and the bottom of the inside diameter of the ferrule (the opening that the operating rod passes through) should be .750 which is 3/4".

J.B. Roberts Jr. had an excellent article in the Rifleman a few years back, in which he stated that a 1" inside diameter ring of 3/16" brass or steel rod could be used for centering these parts, but you had to be very careful to keep everything straight.

I feel that since the density of wood stocks vary so much, along with their springiness, that it was not out of the realm of possibility to take a piece of metal that was 1/8" x 3/4" and use it for a gage fixture that would give close spacing of the two parts. This along with some carefully formed sticks slipped into each side between the stock and the barrel worked, although it took a little filing in the action area of the stock before everything lined up. I later did another M1 that had excellent alignment just the way it came from the factory.

This brings us to another bedding method that was used a long time ago, evidently on some of the first match Garands. It does not appear to be nearly as effective as epoxies but is worth noting, as they did not use any epoxy or glass at all. They simply fitted the wood to the action by filing off the high places, and/or shimming up the low places until the following conditions were met:

1.) The action is firmly bedded into the wood so there is no movement front to back, or side to side.
2.) The action is squarely bedded on the rear raised horseshoe part of the stock.
3.) The front two sides of the action should sit squarely on the wood. (Again, we have the tripod effect.)
4.) The action tang legs should bear evenly against the stock's recoil shoulders.
5.) The lower, horizontal part of the action in front of the front tangs should bear evenly.
6.) The floor plate and rear of the trigger guard must rest squarely on the bottom of the stocks, and there must be a firm clamping action of all top and bottom parts when the trigger guard is locked and closed. This last may be obtained by gluing strips of medium grit emery paper to the low portions of the stock with the grit facing outward.

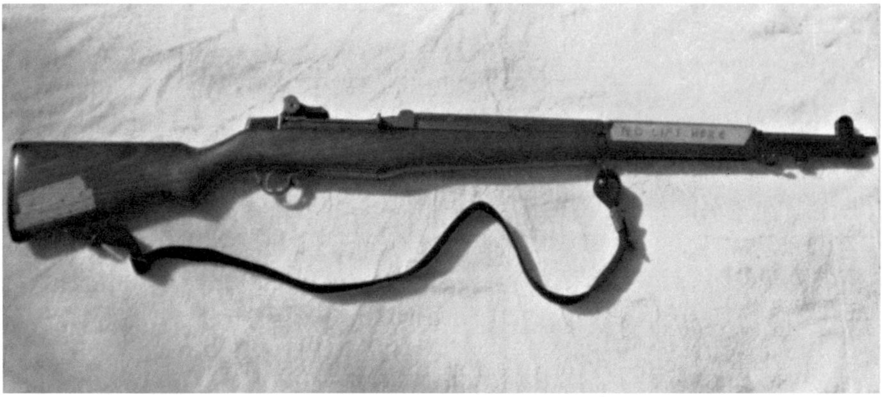

The M1 with its sling. Some like the leather one, some the web, because of its finer adjustment.

7.) The wood must not touch the barrel anywhere.

This method was used during the early days before it became common practice to use fiberglass, and later epoxy bed. It had to be better than nothing, and is interesting as they used the same method that they had used on bolt action rifles for years, using red lipstick, Prussian Blue, and sometimes grease mixed with lamp black, rubbed on the action to show the high places on the wood.

However, time marches on, and in the next part of this series, more tricks on bedding will be discussed. The search for accuracy never stops.

Low Cost Accurizing Tips for the M1

Part 4

More Tricks On Epoxy Bedding

(*Precision Shooting,* April 1994)

As this series of articles on the Garand goes forward, it will deal with reloading tips with GI barrels whose chambers are a bit different than match target chambers; the fit of the operating rod to the barrel, and how critical it is to good accuracy; new developments that have come up on tightening the rear sight to get rid of unwanted lost motion so that its movements will be more precise and repeatable; some tried and true methods of improving the M1, and also the M14/M1A's trigger pull, and what to watch out for; and semi-free floating the hand guards and the gas cylinder. (Editor: All that should take us through the turn of the century.)

One thing that is important is the tension that is required on the trigger guard to lock the action closed after everything has hardened, and again, there are different ways to go about it. It was accepted by everyone (that I talked to) that the best way to get this was that the release

The action goes into the stock in an arc. This is why the action's tangs should have a radius ground or filed on them to avoid gouging the bedding as they are removed or replaced.

agent-coated barreled action and trigger guard were to be assembled into the wet epoxy coated-stock, and the trigger guard then closed until the rear end of it came within about 1/4" of the stock. It is then held in this unlatched position until the epoxy cures.

One way of doing this is to bend a 3/16" diameter rod into a "U" shaped staple until its legs are 1/4" apart. This is then slipped through the hole in the safety, and the trigger guard. Another way is to snap the

Close-up of the radius ground on the bottom rear of the action's tangs.

safety forward, into its unlocked position, and then use different sized number drills slid through the hole in the safety, and the trigger guard. This second method is considered more precise, and is said to give more adjustment for bedding compounds that are thicker or thinner, and rifle stocks that have more or less distance between their top and bottom.

Often a well bedded rifle that does not have the tension desired when purchased "used", can have fresh release agent and epoxy applied just to the bottom of the stock, trigger guard area, and the number drill method used to improve things.

Another trick that some use is to use still more modeling clay, and fill the clip release cutout in the stock and the inletting at the rear of the action mortise and also up at the front, below the action to make room for all of the moving parts when they go back in. Although some prefer to just use release agent here, there is a risk of getting epoxy somewhere that you do not want it. With the clay, it is easier to see, and also very easy to clean out when the time comes to do this.

It is considered standard procedure by everyone to seat the action first, so that you can get the barrel and stock ferrule lined up, then go inside the action and wipe off any excess wet compound that has "squooshed" out. Then fit the floor plate trigger guard assembly. Another thing is to really degrease the inside of the stock and anywhere you want the bedding compound to stick. Any of the good degreasers work for this. But the trick is to be sure and give it plenty of time for the degreasers to evaporate before applying the compound (if you want it to stick). A drying time of a day is best. It is amazing how deep that stuff penetrates into the cells of the wood. But then, it has to, as the coats of lubricating and preservative oils used go partly deep over the years.

After the bedding compound has hardened, and here you want to go with the manufacturers recommended curing time, it's time to take the

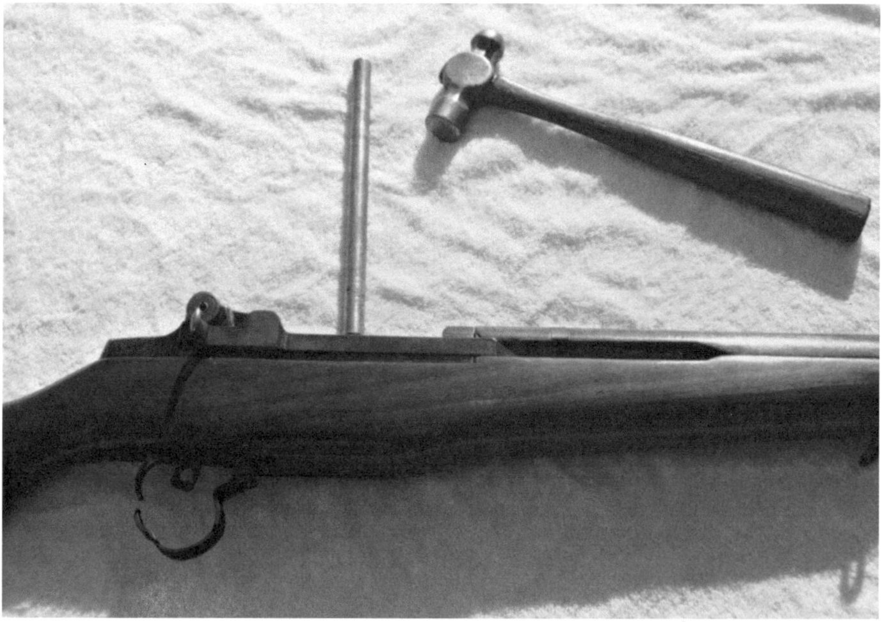

Using a hardwood dowel, tap the floor plate lightly at each of its four corners, and then in the center to walk the floor plate slowly out of the stock.

metal parts out of the stock. Again, there are some tricks. Unlatch the trigger guard and support the stock. One way is to clamp it in one of the cradle vices used for cleaning rifles. This is also a good place to support it during curing.

Insert a hard wood dowel inside the action and tap the floor plate lightly at each of its four corners, and then in the center. If you have been good and not told too may lies, and God loves you and has decided not to punish you for your drinking, repeating this tapping lightly and evenly should walk the guard assembly out nice and straight. Once it is out, turn the rifle over, and start on the action.

Now you have gone to a lot of trouble to bed this thing correctly, and later you should be careful not to shoot overly heavy loads that may pound this bedding loose. So it makes sense at this stage of the game, not to cock or rock the long tangs on the action back and forth as they

The amount of tension on the guard while curling can be controlled by using different-sized number drills through the safety's hole.

slide out, and loosen the whole thing. So you again use the dowel, and tap lightly on the supporting cross-member at the rear of the magazine of the upturned rifle, and then on the ends of each of the four tangs in turn. Again, the idea is to walk the action out slow and straight. It is not a bad idea to have someone support the barrel and action while you do this. Another way, if you don't have a vise, or any friends handy (to serve as operating room assistants to your Dr. Kildare act) is to lay the rifle on its side on a couple of pillows on your bed where it will sink down into the pillows, and tap on it there. This will keep it from sliding all over, and will also cushion the taps. It is surprising how many rifles are taken apart at Perry this way, but if you are at home it probably won't endear you to the heart of the woman of the house. (Editor: Particularly if she was trying to sleep in the bed at the time of this little venture.) But then, accurate rifles are worth some sacrifices, and a good M1 has got to be worth a P.T.O.'ed (partly ticked off) housewife in the long run. However you do it, remember that any time you take it apart, and you never do so unless you absolutely have to, to avoid making it looser, be sure to take the parts out progressively slow and straight. Remember that that long barrel is like a very long handle on a wrench or lever, and if you wobble it around even an inch or so at the muzzle it can easily gouge and loosen your bedding-to-tang fit.

We now come to another decision. Some gunsmiths clean the release agent, or wax, off of the metal parts but not the bedding. Others clean both with degreaser, or hot soapy water, then coat the metal parts with preservative, or lubricant. There is a difference. The old GI cosmolene was a preservative, thoroughly an anti-salt water grease, while the white lubricant, or automobile chassis lube, is a lubricant, but does reasonably well as a preservative. So take a moment to consider what kind of conditions your rifle will be subjected to. It is not uncommon in many states to be caught in a rainstorm at a match that is so heavy that the match has to be halted as the patches get washed off the targets. This is where Mr. Garand's rifle really comes through, as it will work fine even partly submerged, and it is amazing how you can beat people under these conditions that you can't beat in the dry. (Editor: When either I or my rifle are "partly submerged", beating my friends ranks pretty low on my list of priorities of the moment.)

Most gunsmiths felt that if doing well, the bedding-to-metal fit should be so snug that no slipping can take place, and they just clean and lubricate the metal parts. Also, if much of the inside of the inletted area of the stock was degreased, anything not covered with bedding compound should be resealed to keep out moisture which will cause warping of the wood.

So, how much does epoxy bedding help the ordinary GI M1? Most felt that it was the most important single thing that could be done to improve accuracy. Without it, all else was for naught. But if you still have other things that are wrong with your rifle, like a bent operating rod for instance, that is applying pressure or tension to the barrel, you are still not out of the woods. We will deal with this problem later. Each item dealt with should shrink the groups a bit.

But, if most of the other parts on the rifle are reasonably within tolerance, you can expect it to pull the groups down to about 2 to 2½ MOA. Some do better, some do not. But if you were getting 3 MOA (minute of angle) before, this ain't too shabby, and 1½ and 1 MOA are possible with GI barrels, although few will hold this for twenty shots.

If you feel that these MOA are huge compared to match and varmint rifles, you are right, they are! But you are dealing with a rifle that was

The M1 is still used for hunting and plinking. Here my seedy buddy, Hank Cepek, uses his in Wyoming.

designed to win a war first, and a target match second. And reliability under the most horrendous conditions, including air drops, was foremost in its design. The expensive varmint rifle would have been hard pressed to stay with the combat M1 through the mud and snow of World War II and Korea.

Also remember that if you want a really accurate M1, there are some superb gunsmiths advertising in this magazine, and the hard cold fact is that their rifles do win matches.

It is hoped that one or more of the things to be discussed in these pages will unlock the extra accuracy hidden in your rifle, for the search for accuracy never stops.

Low Cost Accurizing Tips for the M1

Part 5

The Operating Rod

(*Precision Shooting,* June 1994)

Sometimes it is hard to say just which has the worst effect on accuracy, a bad barrel on a good M1 rifle or an out-of-alignment operating rod on a good barrel. It is almost like which came first, the chicken or the egg. For no matter how expensive, or how precise the barrel is, if the operating rod, or bolt handle rod, if you wish to call it that, is binding somewhere against the wood of the stock, or bent and applying pressure that dampens the barrel's vibrations, the shot grouping will not be as good as it could be, and there will be flyers! Bear in mind that this also holds true for the M14/M1A as well.

I had read in a book about the M1 that the National Match models had their operating rods gaged, then put on the rifle. So when I went to the National Matches at Perry many years ago, I went to see an armorer

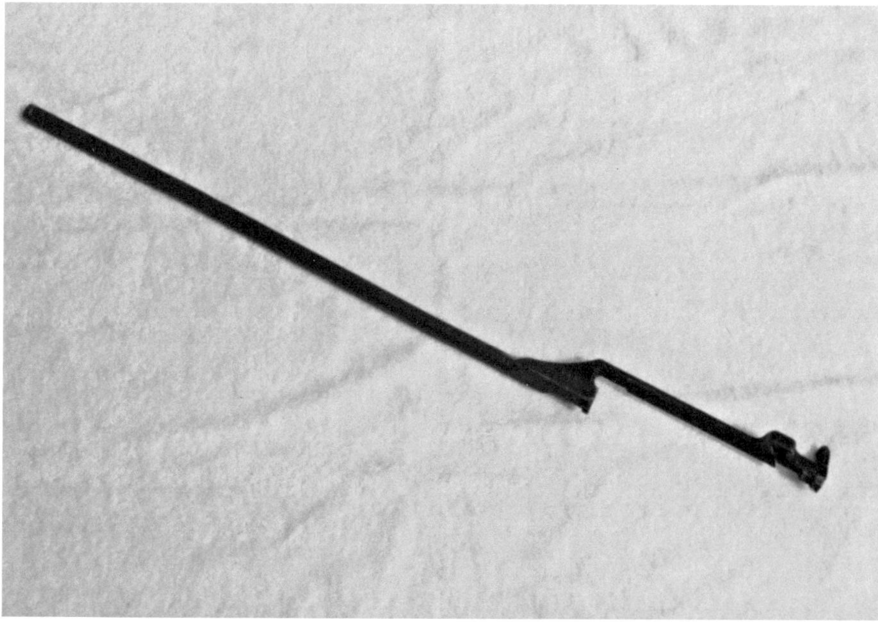

The M1's operating rod. It is not as straight as it first looks.

friend and asked him how I could go about getting this done. His answer surprised me. "We don't fit the rod to a gage anymore." He said, "We fit the rod to the individual rifle." He then proceeded to explain to me how he does it. His method works fine for me, but bear in mind that others may have their own way of doing this. And anything you do, you do at your own risk.

Here are some tests. Test Number One: retract the bolt all the way to the rear. Remove the gas cylinder lock screw, and clean and lightly oil the inside of the gas cylinder, as it must be this way for the following test. I will also mention here that some people like their gas cylinders lightly oiled inside, and some like theirs dry. It all depends on personal preference, and what works best for you. They don't stay dry for very long once you start shooting, as they exhaust to the rear under pressure.

Be sure that the small piston on the end of the operating rod is free of corrosion or carbon build up. Remove the gas cylinder and clean it if you have to, then refit all the parts, and close the bolt.

Hold the rifle at waist level by its pistol grip. Rest the bottom front of the gas cylinder on a waist-high table and grip the front handguard

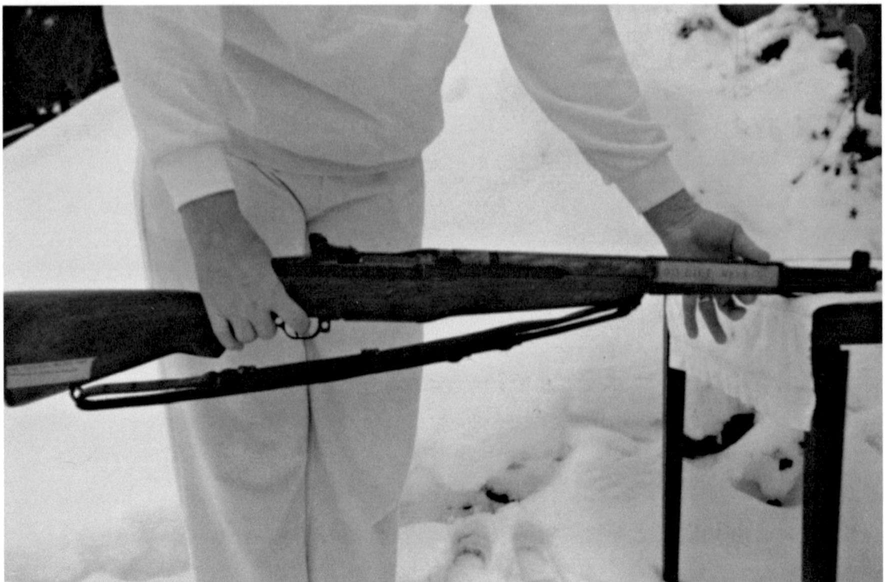

The M1's operating rod is tested for correct movement by holding the rifle horizontal with the bolt closed.

When lifted with one finger, the rod should raise freely to the top of the gas cylinder. when released, it should fall freely to the bottom. This one has its handguard free floated and glued on. Hence the "No Lift Here" sign, to keep from breaking the glued joint. More on this in the future.

with your other hand. Reach under it with one finger and push up on the operating rod, about a half inch behind the rear of the gas cylinder. The rod should raise freely, a tiny amount, until it comes to a stop when it contacts the top inside of the gas cylinder. This would be at the 12 o'clock position if the gas cylinder were observed from the rear. When released, it should fall freely back to rest once again at the 6 o'clock position of the gas cylinder. This concludes test number one.

For test number two, the rifle is stripped to the barrel and receiver group. The operating rod spring and its follower are removed. All of the small magazine parts can be left on as they do not get in the way of the bolt when the rifle is held in the normal top-side-up manner, but you can remove them if you want to. The barrel and receiver group are held sights up, by the gas cylinder in one hand, and with the end of the receiver resting in the palm of the other hand.

The bolt and the operating rod's races should previously be cleaned and lightly oiled to reduce friction. The muzzle is raised approximately

The action should open fully if raised 60° above horizontal. This one opened at less than 40°.

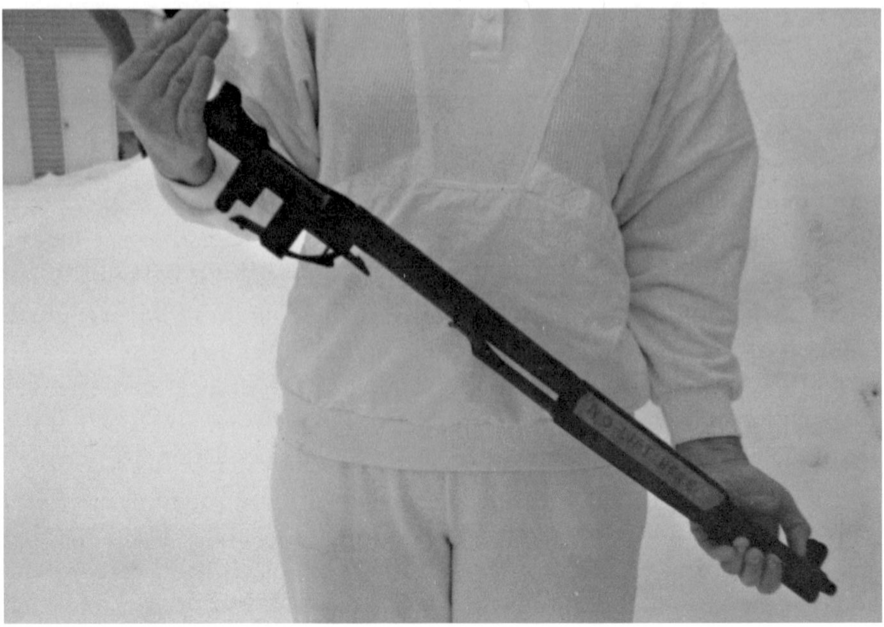

The action should close and fully lock if lowered 60° below horizontal. This one did at less than 40°.

60° above the horizontal. The bolt should open all the way, its full travel, by gravity alone. The muzzle is then lowered approximately 60° below the horizontal. The bolt should travel forward, close and completely lock by gravity alone.

With the rifle held horizontally (normally) with the bolt fully closed, the operating rod may be allowed to lightly rub the bottom of the large diameter part of the barrel just in front of the receiver, and if it does, this area of the barrel should be lightly greased. Opinions differ on this, as it is somewhat similar to epoxy bedding a bolt action rifle's barrel just in front of the receiver ring. On my rifle, I could not get it to meet the above requirements unless it did rub. It must not be allowed to rub anywhere else, especially not the stock ferrule. The ferrule is the metal part at the front of the stock that holds the front sling swivel. This is not to be confused with the stacking swivel which attaches to the bottom of the gas cylinder. Any rubbing usually gives itself away on the operating rod by showing a shiny spot where the phosphate parkerizing has rubbed off.

Evidently, years ago, there were also a lot of opinions on how tight the operating rod's handle should be in its track on the side of the receiver. But most of the people I talked to felt that if the rod met the above requirements, and the handle showed no signs of hopping out of its track while cycling, or being cycled by hand, you were home free.

There were also some discussions on whether or not the operating rod should be allowed to touch lightly against the wood of the main stock, on the right top inside beveled surface of the stock, just forward of the receiver. Years ago, it was thought that this condition was desirable, as it was believed that it dampened the rod's vibrations during the firing cycle, and helped accuracy. However, I could not find anyone who set up an M1 this way now.

There was an individual though, who when questioned about it, thought it was not a good idea to leave anything to chance, and that you could easily put some layers of tape in this area and then try shooting some test groups with ammunition that you knew to give good accuracy, and see if the results were better, worse, or stayed the same. The idea being that if any improvement was found, you could easily replace the temporary tape with a glued-in piece of walnut. Or just remove the tape, if the test was negative.

The bending of an operating rod to make it fit the individual rifle is best left to those with experience, or the bold. The rod, as it comes from the factory has some bends in it already, so that it will fit on the rifle, and if these were completely removed, of course it would not go back on. So when a gunsmith says he has straightened your operating rod, he usually means that he has brought it into blueprint specs, by removing unwanted bends. (He has probably not laid the S.O.B. down flat and hammered it out straight as a frozen black snake as was used in hillbilly tall tales. Since ah had already shot ma ramrod at em... ah grabbed this froze snake... and ah....)

Also, the tubular rod is quite springy, so it has to be bent slightly beyond where you want it, and then it springs back. This alone means that it is a trial and error fitting arrangement, and no one I talked to felt that it was likely to get one spot the first try, as it must be removed from the rifle to be bent.

The procedure used was to look the rod over carefully while it was on the barrel and receiver group, and then decide where to apply pressure to correct any misalignment. It was not uncommon to find that it was out in more than one place, some more than others, and some were quite close.

Many straightened their rods by holding them in a wood padded vise, and applying pressure with their hands. But this was dictated by where the correction was needed, so some used the vise and grooved blocks. One used screw eyes and clamps fitted to the work bench, with a twelve inch rule at each end, so that he could bend the rod so many inches, and then let it spring back so many inches, until it came out just right. The slickest I saw was a set up on a Bridgeport mill table using blocks and the electronic digital readout of the table's travel.

I found that mine was rubbing the inside of the gas cylinder at the 9 o'clock position. I took it off the rifle and put it in the wood jaws of a vise and kept trying until it came in. Another thing I found helpful was to remove the gas cylinder lock screw and lock, and while looking in from the front, with a magnifying glass, shine a small flashlight along the rod from the rear of the cylinder. In this way, I could more easily see the alignment of the rod to the cylinder.

How much the groups shrink on a given rifle after the operating rod is fitted correctly depends mostly on how much it was interfering with the barrel's natural vibrations previously. Sometimes it just eliminates flyers, which is something to thank the Lord for. But most felt that on a standard GI M1, that fired National Match ammunition into about 4 MOA., that epoxy bedding it should pull it down about 3 MOA. and aligning the operating rod should bring it down close to 2 MOA. Of course, you have to understand that these are very general terms, and one rifle may show a very considerable improvement, and another not much at all if everything was pretty close to specs to begin with. Mine did about as described above.

There are an awful lot of M1's out there. There were over four million manufactured by Springfield Armory (the government arsenal, not the privately owned company), Harrington Richardson, Winchester and International Harvester from about 1937 to 1945 alone, with the M14 coming in about 1957. Add to this that many consider highpower target shooting to be the NRA's premier discipline, with about 25,000 classified competitors in the bolt and service rifle classes and that this number is growing every day as more M1's are sold through the DCM, and you begin to see what a far ranging program the M1 is involved in.

No discussion on the M1's operating rod would be complete without a warning to stay away from very slow burning powders like 4831 and 4350 when reloading.

The M1 was designed to be phased in gradually to replace the 1903 Springfield. (WWII of course speeded this up considerably.) So it was made to use the 30-06 ammunition already in use. There had been some consideration to change the caliber to 7mm. But this was given up and we stayed with the 150 grain ball and 173 grain match rounds.

So John Garand tapped the barrel to get the gas pressure to work the action, right up at the muzzle, where the 50,000 or 50 P.S.I. breech pressure had dropped to about 6,000 (+ or – about 2,000) just before the bullet left the barrel.

If you reload with a powder much slower than 4320, the remaining pressure will be too high when it enters the gas cylinder, and sooner or later, this will bend the long tubular operation rod just as if you stood it on its end and continually hit it with a big hammer. Then you could be right back where you started from, and possibly even worse off, depending on where the bend or bends occur, and what effect these have on your barrel.

The next part of this series will deal with some of the reloading methods used by top shooters, gunsmiths, and armorers. Over the years, they have come up with some nifty ideas. The search for accuracy never stops.

Low Cost Accurizing Tips for the M1

Part 6

Reloading

(*Precision Shooting,* August 1994)

When you first handle an M1 you think, ah, another 30-06... only this one is an autoloader. In some respects, this is true. But it is not "just another 06". It has some small problems that go with its design, and it took a lot of gunsmiths, armorers, and experimenters a long time to learn to make these things work for them. They have been good enough to pass these tips on to us so that we could pass them on to others. Our hats go off to these people, for they are the best.

To begin with, the M1 differs from a bolt action target rifle in that it is designed first to be a military rifle, and must be able to withstand tremendous abuse. If you were to suggest to George Kelbly or Walt Berger that they take the fine precision benchrest rifles that they used to win with in the World Championships at Finland, and follow you with your M1 through an air drop and a swim through a muddy river before starting to shoot, they would probably explain to you nicely that their rifles were not designed to do this. They would probably also like to stick your ear in the chuck of their lathe and tighten the handle a couple of turns. But you begin to get the idea. As such, the service rifle usually competes in its own class. But a good one, with a good shooter behind it, can and does finish ahead of a lot of bolt rifles in highpower matches.

At the start, the M1 differs from its bolt action match counterpart by having a barrel that is much thinner from the forearm out. It differs also from the light barreled sporter by having the unequal weight of the handguards, gas cylinder, and long operating rod with its spring, hung onto its barrel. All of this is then connected to the lower main stock by its bedding, and the barrel's lower band mating to the stock ferrule. It is hard to imagine a much worse arrangement for precision accuracy, especially at extended ranges. In spite of all of this, the rifle does quite well, often turning in excellent scores.

A match bolt rifle is designed first and foremost to win matches. Everything else takes a back seat. You have only to look at what we do to their action's stocks and barrels to see this. The hunting rifle is designed to do just that, hunt.

The service rifle is designed around much different perimeters. It must first be able to survive all weather conditions, most abuse, and still be totally reliable, win battles, and bring you home alive. After all of that, it is expected to do well in target matches. We have special match ammunition to help it do this, and it will.

This match ammunition does very well in the standard GI barrel, and it naturally does better in a match barrel. The standard G. I. chamber may well be cut bigger than its match counterpart so that it can handle ammunition that is bent, dirty, dented, etc. Because of this, small base sizing dies are not usually considered necessary and many people do not resize the case any more than is needed to make it fit the chamber it was fired in. Also, as mentioned in Part 2 of this series, it is very important that the muzzle's factory countersink crown be checked for accuracy, and redone if necessary, as it may well have been put on in haste.

The GI barrel on a used M1 must be considered an unknown until proven otherwise. For though some rifles are issued over and over again, and seldom fired, this does not mean that the muzzle crown does not get nicked, because it often does. If the rifle was fired, it should have been cleaned correctly, but that does not mean that it was. There is also the other end of the scale in which rifles used at training camps get thousands of rounds put through them. Rifles from the DCM are normally refitted with whatever they need before leaving, and I have seen many that were absolutely outstanding. Still, there are others from other places, and you just don't know how many rounds went through them.

Because of this, unless you know for a fact otherwise, every bore should start out with a thorough cleaning, and if necessary (only if necessary), consider an abrasive cleaner such as Rem Clean. If you do decide to go with an abrasive cleaner, turn the rifle sights down when cleaning so that no abrasive will enter the gas port. Considering the Garand's history, it is entirely possible that one could be out there with thirty or more years OF built-up fouling.

Once the bore is considered O.K., most Garands will be found to do very well, or even best, with either the Government National Match load or its reload equivalent, which the American Rifleman Magazine lists as 46 grains of IMR 4895 and the government M-72, 173 gr. match bullet. This falls right in line with Lyman's reloading manual which uses the same powder, but a 180 gr. bullet seated 3.280 o.a.l. The government round is usually 3.320 o.a.l. (overall length). Naturally you should start below this and work up slowly, not only for safety's sake, but because you could well find a slightly reduced load that is a world beater in your individual rifle. It has happened more than once! Generally speaking, the government did a whole lot of testing before it arrived at its match load. Some handloads will beat it in some rifles, and some will not. Also, because the throats of GI chambers are longer than match ones, some find that experimenting with longer overall length (seating depth) can yield significant improvement. But again, you have to watch out for pressure rises, and progress slowly.

Some 168 grain loads do fine. Still, others find that a 180 grain flat base hunting bullet will perform better in their rifles than any of the match bullets. The old story of some barrels just plain not liking boat-tails is very real, and I have seen some outstanding groups fired with flat base bullets and starting loads of 4320 powder.

Speaking of powders, it is important to know that the Garand was designed with powders with a burning rate not faster than IMR 3031, and not slower than IMR 4320. Burning rate, or relative quickness of burning, gets pretty technical because there are so many things that can change how fast the powder burns. Almost any increase in the amount of pressure it takes to make the bullet move can make the burn rate rise rapidly. Generally speaking, if one powder requires more grains than another powder to get the same velocity and pressure, it is said to be slower burning.

If you try to use powders that are too fast, you can run into high chamber pressures, but still not have enough pressure at the gas port to cycle the action correctly. Powders that are too slow can give port pressures that are way too high and their sharp impact will sooner or later bend your operating rod, all the while giving chamber pressures that are well within the safety range of the rifle.

Another thing that can happen with the wrong powder and gas port pressure is odd feeding failures. With too fast a powder, giving low port pressure, the bolt can fail to travel far enough to the rear to let the next cartridge in the magazine rise in front of it to be fed. This is called "short cycling". Or, the bolt may feed the cartridges from the magazine, but not travel far enough rearward to lock open on the last shot or eject the empty clip. At the other end of the pressure curve, too slow a powder, giving too high a port pressure, can cause the bolt to cycle so fast that the cartridges in the magazine do not have time to rise up and be fed — especially if only one or two rounds are left in the magazine, and the magazine's spring is getting lighter.

I picked up some VihtaVuori N 160 powder a while back, and it was doing so well in my bolt action 06, that I wondered how it would do in the M1. But I noticed that their burning rate chart listed it as right next to IMR 4831. So I called their sales agent, Ross Duckworth. We had met Ross at the Nationals at Perry when he was talking with Marcus Raab and Allen Ewing of the U.S.A.R. team. Allen, by the way, won the Porter trophy 1000 yards service rifle match, with Marcus coming in third behind Grant Singley. The other 1000 yard service rifle match is the Farr trophy, won by Dave Kerin. These matches are fired concurrently with the Leech and Wimbledon. This was in 1993.

Ross also felt that this was too slow for the M1 so I shelved the project. He is pretty sharp on this stuff. I intended to try N135 later.

A well lubricated Garand should feed flawlessly. I like the government's idea of using white lubriplate grease because as soon as it is dirty from firing or other contamination, you can see it getting darker and change it. This works well on the M1A too, but it is also a good idea to put a few drops of an oil like Brake Free on the axle of the M1A's bolt cam roller.

If you do not have a copy of the M1's field manual FM 23-5, you should get one, as this covers all of the little things that are so easy to overlook on the care, cleaning, and lubricating of the rifle. You can get these at gun shows, parts and book dealers, and I think the DCM The short version is cheap.

One of the standard issue oiler-cleaning units is nice to have along at the range. More than once, I have seen the steel GI. cleaning rod loaned to remove a stuck case in a bolt rifle, so an expensive rod would not be damaged or bent. Also, I have found that lubriplate works so well, that I went to the local bearing store and got a 5 lb. can of it to use on restored classic and sports cars.

FM 23-5 states that the top rounds of the M1's 8 round clips were loaded at the arsenals to be on the right-hand side of the clip, for ease of loading using the thumb of a right-handed firer. I know left-handed shooters who put the top round on the left for the same reason. Also, some M1's may suffer what is known as the "7th Round Jam". When this happens, the 7th round may smoke stack, or hop right out of the rifle. There can be a couple of reasons for this, and each has its cure. First, make sure that you are using ammunition that does not cycle the action too slow or too fast. This problem almost never occurs with GI ammo. If, however, you have one of the rare M1's made just before WWII that had some manufacturing shortcuts, try loading the top round on the left. Also, any new or rebuilt M1 should have its action parts lubricated (and kept that way) and may require 25 to 100 rounds through it before it wears in, and wears some of its phosphate coating off, before it can be considered 100% reliable, and then it will be just that, 100% reliable.

No discussion on reloading for the M1 is complete without a warning on slam fires. It is believed that there are several things that can cause this (the firing if the rifle before the bolt is completely locked closed), and it is almost unheard of with a GI rifle using GI ammo. Most believe the biggest offender by far is the reloader failing to completely clean the primer pockets and failure to fully seat the primer. This is easy to do when loading large quantities, or just plain not being careful. Beware of distractions. It also helps to place cases in loading blocks as they are primed instead of tossing in a box. Then if you wish to go back and check one, you know right where it is.

Rule #1 with any auto loader, and especially the M1's, is "NEVER LET THE BOLT SLAM CLOSED WHEN LOADING SINGLE SHOTS." Yes, there have been slamfires when feeding from the magazine. But it is believed that the act of the bolt stripping a round from the magazine slows the bolt's closing speed, and puts the odds in your favor. Also, a resized case does not have to be sloppy, but it must be able to go into the chamber freely, as it appears that a high case plus a high primer can also give trouble. An 8 round clip, modified to stay in the rifle, is very handy for firing, single loading, and is especially helpful when firing left-handed, or prone with a sling.

Treated right, the M1 is a superb rifle. Part 7 of these articles will deal with more on reloading, for the search for accuracy never stops.

Low Cost Accurizing Tips for the M1

Part 7

More on Reloading

(*Precision Shooting,* October 1994)

It appears that there is no end to the tricks of reloading, or the theories that lead to them. Some of these are pretty far out in left field, and some of them are so simple, and yet effective, it is surprising that they have been overlooked.

Quantities of M72 30-06 Government Match ammunition can still be found from time to time. The 173 grain Boattailed Match Bullet used in this round has a black tar-like sealer used between it and the case neck to protect the charge from moisture during storage, which can be for long periods of time. It is amazing how much of this good stuff was shipped to Vietnam during the war years.

One of the tricks old Perry shooters used to do with this fine cartridge to enhance accuracy was, a few days before a match, run the rounds to be fired through a bullet seater set to push the bullet about .010 to .020 deeper into the case, thereby breaking this seal and making the bullet

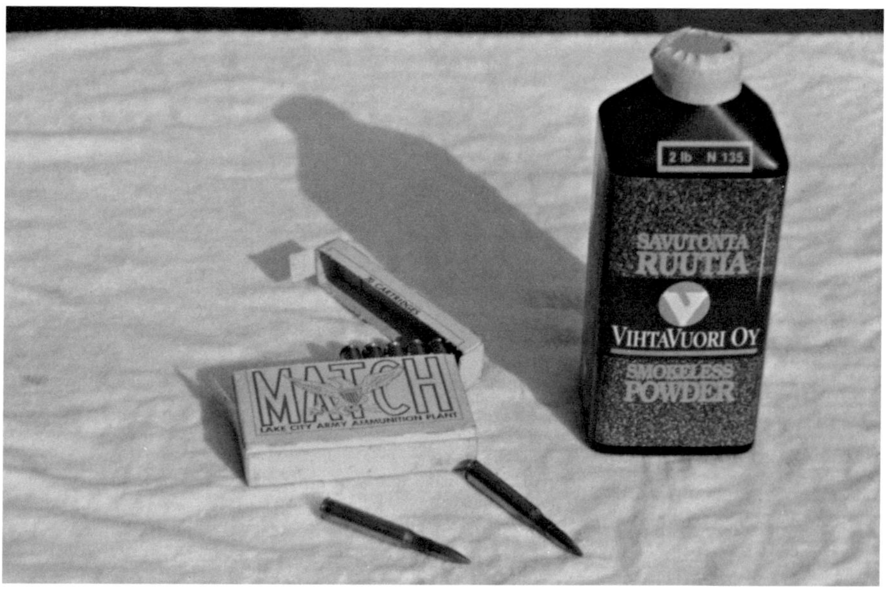

N135 "V" powder works very well for M1 match rounds.

pull more uniform from round to round. It was very common to log the date at which this seal was broken on the side of the cartridge box for future reference, in case they remained unfired for some unforeseen reason. As a matter of fact, I have a few boxes of this, so marked, sitting on the shelf by my loading bench right now.

A new wrinkle to this trick was shown to me by another shooter this past summer. Believing that even though the seal was broken, and that things now should be better, the thought occurred to him that it would be highly unlikely that the coating would be as uniform as he liked around all surfaces of the bullet. He also did not like the thought of this

A season's shooting of ammunition can be seated long, and stored in .50 caliber ammo boxes in the same area that you store your powder.

stuff in the lands and grooves of his fine barrel. With this in mind, he set the cartridges in a loading block, pulled all their bullets and looked for some way to clean them. The first thing he tried was sloshing them in a cake pan or gasoline, and this worked like a charm, the gasoline turning black, and the bullets coming out shiny clean. He then reseated them, only this time about .030 longer in overall length. It worked for him. I just wonder how it would work with 7.62 match ammo? By the way, he used a Forster collet-type puller in his press, and always broke the seal first, to ease the pulling.

There is still another trick in regards to loading quantities for the coming season. It is not uncommon to size, trim, and prime enough cases in the off season, so that they can be quickly loaded as needed for matches as the season progresses. It is felt that in this way, the ammunition would be more fresh than if fully loaded in the off season and then stored. The trick now is to fully load when not busy, but only seat the bullet to 1/2 of its normal seating depth; box and date these rounds; then store them in .50 caliber metal ammo boxes, that have good seals, in the same conditions, or place, that you normally store your powder. Then, a few weeks before a match, final seat what is needed. In effect, you have stored the powder in measured charges, with the cartridge case becoming the container, and the bullet the plug, or cap for the container.

There are yet a couple of more wrinkles to this trick. One is to place a small mark on the side of the case facing you with a felt pen as you half seat the bullets the first time. Then it can be placed in the press facing away from you when final seated. Depending on the press, this can minimize press induced bullet top runout. But it does not hurt to check this out with a dial indicator before getting too involved, for your case may well be different.

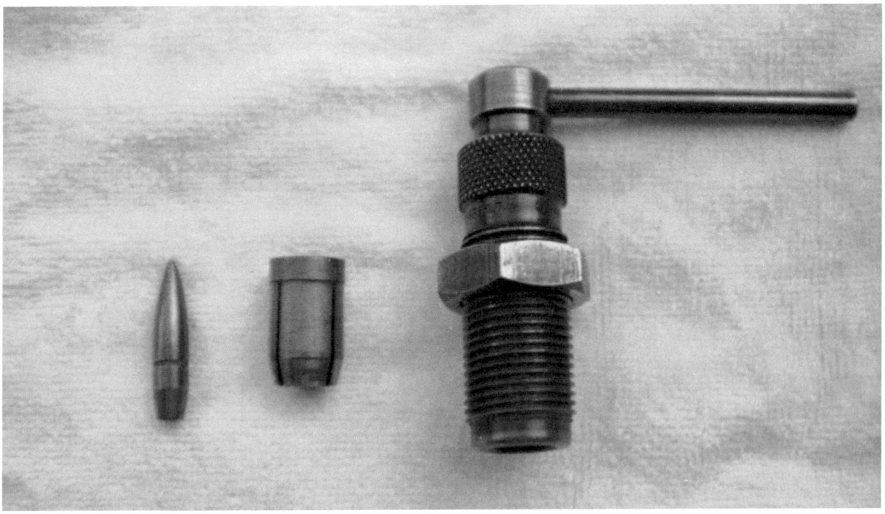

A Forster collet-type puller is hard to beat for changing bullets in assembled loads.

The other beauty of this method is that if you decide to change bullets or powder charges for different ranges, or because you found a better load later, it is a simple matter to pull a 1/2 seated bullet using a collet-type puller in your press. Both Forster and R.C.B.S. make them.

Still another of my oddball friends (and I have many of them, I am happy to say) likes his cases very bright and shiny so that he can find them where they hide in the grass during the time allowed for a brass pick up call. I keep telling him that we could make a fortune training a retriever dog to do this. Perhaps a near-sighted Brittany or an illiterate coon hound (can't read coon sign) or maybe even a pro gun chimpanzee. Anything smarter than the rest of our friends on the line.

Anyway, what he does is size and trim a huge number of cases, but does not deprime them. Then, a few weeks before a match, he runs them

It is not uncommon for match shooters to have boxes of National Match ammunition sitting on the shelf with the date at which its seal was broken logged on it.

through his case polisher, then deprimes them, cleans the primer pockets, and loads in the normal way. Using this method, he never has to worry about any polishing grit trapped in the case's flash hole causing ignition problems, as any there would be knocked out by the depriming rod. For years, he used a cheap #4 drill, ground to a flat bottom configuration on the side of an emery wheel, then held horizontally in the jaws of a vice, to clean the primer pockets with a 3/4 turn of his fingers. He now uses one of the slick pocket cleaners from Whitetail Engineering.

The Garand is very friendly to left-handed shooters, the sight knobs and safety being for all practical purposes, ambidextrous. There is,

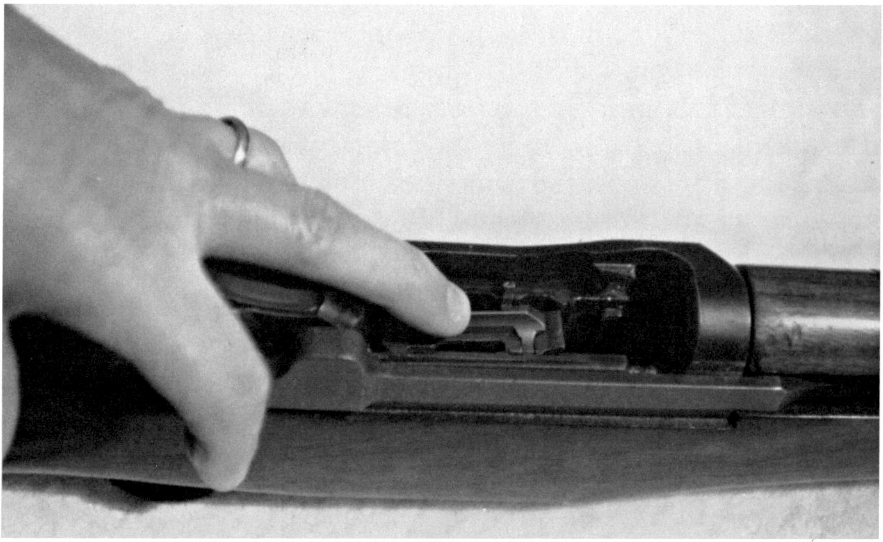

When closing the M1's bolt left-handed, the heel of the hand is cupped firmly over the sight, and the magazine follower is de-pressed with the middle finger.

however, a slick way for left handers to close the bolt that is not generally known. This was shown to me years ago by a friend at Perry who is now with the Border Patrol. It goes like this....

The heel of the left hand is cupped over the rear sight firmly, so that it cannot be drawn forward with the bolt. The key here is firmly. The longest finger of that hand then depresses the magazine follower, and either the first finger, or the thumb, whichever works best, is used to slightly retract the operating rod (bolt) handle. This finger (or thumb) then lets the bolt travel forward slowly as far as it can. It then releases the bolt for the remainder of its travel. If the hand is firmly cupped over the hump of the rear sight, pressing downward against the rifle, the nose of the bolt should lift the fingers out of its way as it travels out from under them. The only way an M1 finger can be gotten using this method is if the left hand is allowed to move forward with the bolt. Naturally, it takes some practice to learn to do this safely, as does the right-handed method, but there are some who believe that this way is safer than the military's right-handed way. It is easy to pick out the ones who need more practice, as they usually grip the afflicted hand with the other one, bend over, and emit sounds somewhat similar to a call to prayer. Like anything of this nature, you assume all risks. I have also seen this used on the M1A/M14, but it is easier with this rifle to just drop the magazine, retract the bolt, and close in the normal manner. Most people tend to use both hands to close a bolt on an M1 anyway, but it is nice to know that there is an alternative to the military method.

There is also a trick, sometimes called the double click method, of unloading an M1 with a partially filled magazine. With the safety on, it goes like this. The right hand is cupped over the action, and the bolt handle is pulled all the way to the rear with the edge of the right hand. While holding it there, the clip release is squeezed with the thumb. This catches any live round that might have been in the chamber and lets the clip rise up out of the action. How high it will rise depends on how many rounds remain in it, and how deeply your hand is cupped over it. If your hand is cupped high, you risk losing the round from the chamber, if it

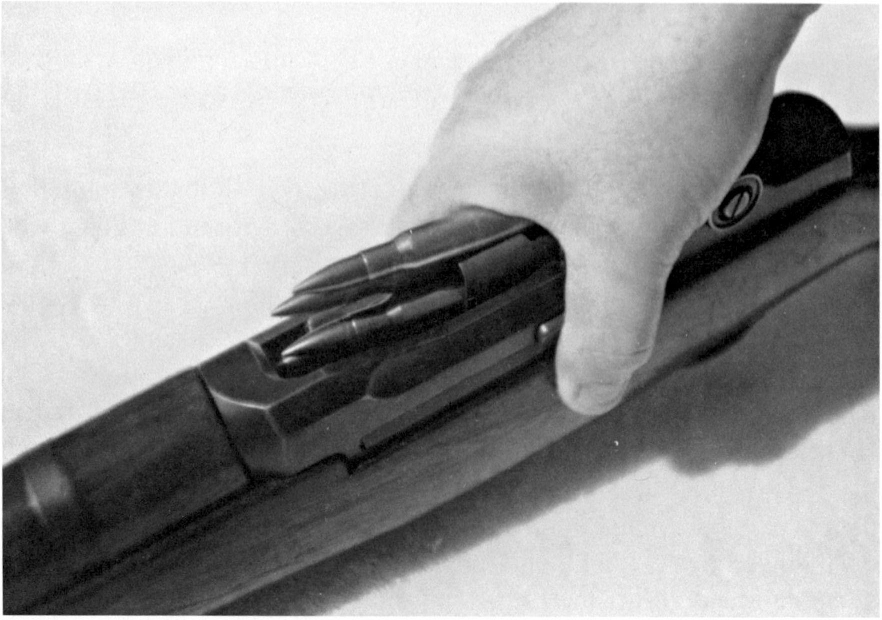

With the double click method of unloading the M1, the hand is cupped deeply over the rounds before the clip release is pushed.

A young student at the DCM clinic learns to close his M1 rifle.

And... he gets it closed.

is low the clip cannot rise high enough to let the follower parts lock the operating rod (bolt handle) to the rear. If it is not locked to the rear, removing the clip frees it to fly forward. The simple solution, that seems to work best for most people, is to cup deeply, pull the bolt to the rear, squeeze the clip release, and then let the bolt come forward until it stops against the back of the raised clip. Then, still cupped, pull it back a second time, and it should lock then. By cupping the hand deeply, you contain the chamber round as it is released by the powerful ejector. The second pull frees the clip so that the follower can shove it up into your cupped hand, with the round from the chamber neatly trapped on top of it. This may sound like standard unloading procedure to some, but where the difference comes in is in how many rounds are remaining in the magazine. If only 2 or 3, the clip and follower will probably have enough room with the average hand to rise and lock on the first pull. But with 4 or more, it will often require the second pull. This is another one of those things that sounds so complicated at first, until you have tried it a few times and seen how simple it is. For some, it saves the embarrassment of cartridges strewn all over the ground. I have also seen this operation performed satisfactorily left-handed.

Another thing to remember is that any handload to be used at 200, 300, or 600 yards should be worked up at the distance it is to be used at if at all possible. Rationally speaking, if a load does not perform well at 200, it usually will not perform better farther out, and a load that looks great at 100 may well fool you and do poorly at 300, and 600. I have gotten snookered on this more than once. But sometimes a load that will hold the 10 ring, but not the x ring, at 200 will continue to hold it right on out. For many people this is fine, as some percentage of those shots will land in the x ring, but for others this is still not fine enough. I have seen many loads that would print into 1" to 1½" at 100 that would not hold the 6 foot by 6 foot target frame at 1000, and so have plenty of other people. Long Range High Master shooter Larry Moore once told me that the only way you could tell if a load and rifle were going to be any good at 1000 yards was to shoot it at 1000 yards. The same goes for the closer ranges.

A good rule of thumb for a highpower shooter is to start the testing at 200, and any load that looks good should be tested on out, and considered for use at where it does its best, for the search for accuracy never stops.

Low Cost Accurizing Tips for the M1

Part 8

The Rear Sight

(*Precision Shooting,* December 1994)

In April of 1988, American Rifleman Magazine featured an article on improving and tightening the rear sight of an M1, M14 by epoxying, plus using shims and springs, that was excellent. No, it was better than that, it was superb. That article was done by one Buddy R. Salyer, a talented writer who has several times appeared in these pages. If you can find a copy of this well thought out article, by all means get it, for it shows one way to go about fixing the sight's looseness. There are, however, other ways, and like anything else, we find some armorers and gunsmiths who approach the problem from a different direction. This is as it should be, as it lets each of us decide which way we want to go, and does not take a thing away from the excellent work done by Mr. Salyer.

The rear sight on Mr. Garand's rifle is without a doubt the finest ever to appear on a combat service rifle, and I also rank the M16 A2 sight up there with it. However, in order to interchange with all the other sights out there in the field, and still be able to work when submerged in mud or snow, it must, of necessity, be made with its parts fitted together with

The M1 sight disassembly. It is really quite simple.

generous clearance. Even though it is well spring loaded, the average GI sight sometimes does not repeat well, and some can give flyers. When this happens, the shooter has no way of knowing what was at fault: bad operating rod, bad ammo, bad barrel, bad wind doping, or bad shooter, etc.

The simplest way to cure some of this is to snap a common rubber band over the side of the aperture and under the elevation knob. This method applies pressure both downward and to one side of the aperture,

The time honored trick of snapping a rubber band around the aperture and under the elevation knob is cheap insurance on any M1 sight, as it takes up any elevation and windage free play in one direction.

tending to take up the free play of both the elevation and the windage in one direction, leaving the windage knob free for quick corrections, and is very cheap insurance on even a good tight sight in competition.

The sight is easy to take apart into its five basic parts, which are the windage knob, cover, base, aperture, and elevation knob.

The cover is actually a bent piece of spring sheet steel which applies downward pressure on the aperture and base to take up the free play

(Mr. Garand was a very smart engineer). It does this through the 1/16" x 5/8" horizontal detent groove which is right in front of the aperture, pressing on it.

The single piece aperture incorporates not only the sighting hole that we look through, but also the curved pinion teeth of the rack and pinion type of elevation mechanism... the ring gear being on the elevation knob shaft.

Hooded match apertures with different diameter eye holes (apertures) are available, with the center of the hole offset .004" so that they can be rotated to give 1/2 minute of angle changes in elevation. They can also be used to get 1/2 minute windage changes if you pay close attention to the location of the small "V" notch cut in the top of the hood, and set it to either the three o'clock or nine o'clock position. But they won't click lock here!

There is also the slick method of filling the eye hole of the standard GI aperture with a slightly harder than walnut wood epoxy such as micro bed, and then putting any size hole in it that you want by turning number size drills through it with your fingers. The beauty of this is that

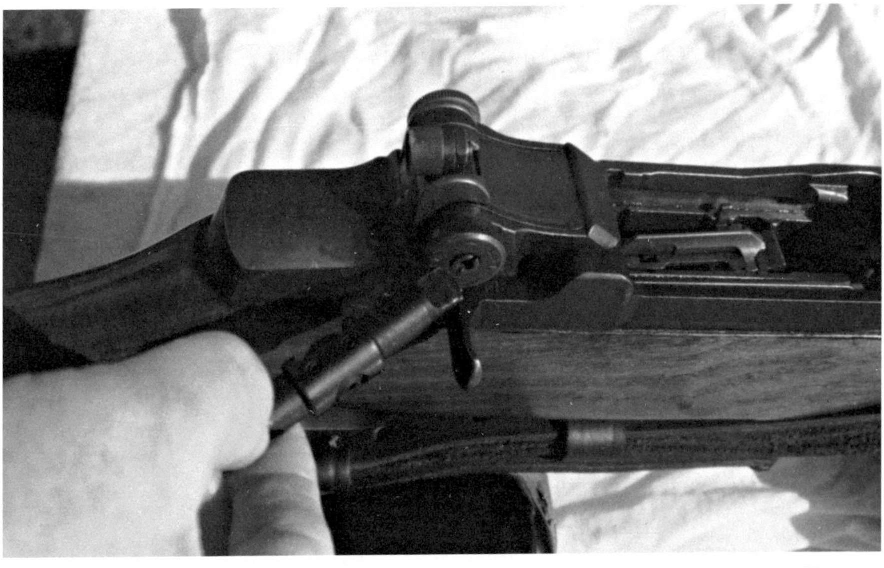

The M1's screwdriver on the end of the combination tool is excellent for adjusting the spring tension of the elevation and windage knobs. This is done one click at a time through the screwdriver slot in the center of the windage knob.

if you goof up the venture, it's easy to do over, or remove.

To disassemble the sight, the screw in the center of the windage knob is loosened. It is spring click adjustable (more on this later) and stays in the windage knob. Your fingers are all that are normally used to hold the knob from turning while the screw is loosened. Not surprisingly, the screwdriver that comes on the GI combination tool is perfect for the job. With the screw loose, the windage knob turns right out of the sight. The elevation knob and its shaft are then removed from the other side, and the aperture is slid up out of the base... the base being the part with the big ear type guards on each side of the aperture.

The base is held into the receiver by the ends of the sheet metal cover snapping into slots in the receiver, and is removed by pulling upwards on the base's ears. It is replaced by pushing in and downward at the rear of the base. When removed, it makes a snapping noise that is exactly like that of a very expensive Olympic micrometer sight being broken, so don't be surprised if your heart skips a few minutes worth of beats. The aftermath is somewhat similar to having stepped on your neighbor's cat in the darkness.

Once all of the parts are out and cleaned, it is easy to see how loose fitting they are. The aperture-to-base fit is one of the biggest offenders. This can be lessened by Buddy's method of using epoxy, or another

One way to decrease the clearance between the sight's aperture and base is to anneal the base and press the parts together.

way that is used by armorers, in which the hardened base is heated until red and allowed to cool very slowly. This is called annealing, or softening. The aperture is then placed back in the base, and the base is squeezed in a vase to press it against the aperture. If squeezed overly tight, the aperture is tapped out with a brass punch, lightly coated with a fine grit lapping compound, such as used to finish-lap automobile engine valves, and passed back through a few times until a no wiggle, sliding fit is obtained. Though crude, this method has a lot in common with the fitting of a match target slide to an accurized .45 pistol, and it does work! As time passes and the sight is used, it is a simple matter to re-squeeze again to compensate for wear. But it is very important that the base be annealed before squeezing, as they are very hard and brittle, and will break if not annealed. However, they are not a high cost item, and can be purchased from most M1 parts dealers.

A couple of other things happen when this method is used; both of which are more or less beneficial, depending on how loose the individual sight is, and where this looseness occurs.

The first result is that if the base were viewed from end on, it would now be found to be bowed slightly. This causes its bottom to protrude downward slightly in relation with its axis, which is the elevation shaft, and this tends to take up some or, in some cases, all of the free up-and-down play that is usually under the base. In Buddy's article he measured the free play under the base with a feeler gage, and then fitted a brass shim of correct thickness. In some cases, GI sights being different from each other, this may have to be done for best results, even though the squeezing method was used.

Bear in mind that to measure this bowing correctly would require some precision instruments, but the difference is there, and it is in our favor.

The second result that takes place when we bow the base is that the windage knob threads in the base are now bent slightly off center-line from the mating threads on the outside of the windage knob shaft. What this does is force the loose GI thread all to one side, and since the threads are "V" shaped, it in effect forces the "V" shape of one into the "V" shape of the other, which tends to make a screw-to-nut fit with almost no backlash, which of course is just what we want. Buddy used a coil spring to control this windage lost motion, and since sights are sights, some may find that they need this too. The dimensions given for the spring in Buddy's article are 3/4" long, .350 outside diameter, and .030 wire thickness.

I also know of people who make up a no backlash thread by coating the inside threads with a steel based epoxy and coating the male outside threads with release agent; they then screw them together and let them cure. I can assure you that this is for those who know exactly what they are doing only, (for instance... it is not the sort of job that our beloved editor should even think about tackling) for these are very fine threads, and axles are not designed for large foot pounds of torque that go with a screw-up.

Once the sight is finished, all parts that are going to rub against each other should be lightly greased. Some armorers prefer a heavy grease rather than a light oil so that if there is any lost motion left, especially when the sight is cranked up to a high elevation where the aperture has far less support, the heavy grease will allow less free movement through recoil than the light oil will, through more resistance.

Two more things can be done once the sight is assembled, which of course is the reverse of disassembly. One is to count the number of clicks up from the bottom to your 200 yard zero, using the load that you like best, or the national match load, or its equivalent, which is very close to Lyman's 46 grains of IMR 4895 with a 168 gr. Sierra Match Bullet. Then set the elevation dial on 200 by holding the knob with padded pliers and loosening the locking screw in the center of the knob. By applying sideways pressure on the aperture with your thumb, you can hold it in place while the dial is set to 200 and then with the knob again held, and the lock screw is re-tightened. It is nice to be able to set it on 200 to get 200, and 300 to get 300, etc.

The other step is to correctly adjust the spring tension that regulates the sight's elevation and windage. To do this, the sight's aperture is raised to its full height, it is then lowered two clicks and the rifle held by its pistol grip. With this same hand, the thumb is placed over the top of the aperture and a downward pressure applied. If the aperture drops, the spring tension must be increased by turning the screw in the center of the windage knob, one click at time, in a clockwise direction until it does not drop.

If it does not drop when thumb pressure is applied, the adjustment is either correct, or too tight. If it is too tight, the windage knob will be harder to turn. This will be very noticeable when trying to rotate it while

Correct spring tension on the M1 sight is checked by pressing downward with the thumb.

prone, or with sweaty hands, left-handed, by a person of small stature, with gloves on, or with the bolt retracted and there is less room for the hand because the operating rod handle is in the way. This can be extremely frustrating during a match. The cure for this is to back off the spring tension screw one click at a time until the aperture will drop from thumb pressure, then turn it in again until it does not. Proper lubrication here makes all the difference in the world. When set up correctly, the sight will not move from recoil during long strings of firing, but the windage knob will still be easy to turn.

It takes the new shooter a little while to learn that it only requires a small amount of movement on the elevation knob to get a one minute of angle change, while the windage knob requires a quarter of a turn to get the same amount (on a standard sight) and this is best learned through doing, and lots of practice.

The search for accuracy never stops.

Low Cost Accurizing Tips for the M1

Part 9

The Trigger

(*Precision Shooting,* February 1995)

If we take a bunch of shooters who hold an expert classification using bolt action match rifles, and maybe even throw in a master or two, and put them all on the firing line with a really top flight service rifle shooter, he (or she) will come out the winner almost every time. Well sure, you think, they hold a higher classification. But wait a minute,

An M1 and a good dog are nice to have while hiking. Here a hillbilly housewife (mine) takes both of them along.

they are also shooting a rifle with a minimum of a 4½ pound trigger pull! So how can this be? How can they compete against our 2 (and less) pound pull bolt rifles? The answer of course is that they have learned to live with it. If you ever work on race cars, your world becomes centered around lap times, and you soon learn that the biggest variable that you can introduce into the car, with all of its adjustments, is the driver. The very same holds true with target shooting.

A top service rifle shooter mentally learns to develop a trigger release that is best described as "somewhat different". Since the service rifle's trigger is two stage, they first take up the slack of the first stage when the sights are centered on their target. Then some use a sharp, fast release like an offhand Olympic shooter. Some might even compare this to that of a benchrest shooter. While others use a smooth, straight back, quick movement, rather like that of a double action target revolver shooter. This type of release is now being taught more and more as double action autos gain in action shooting competitions.

All of these releases continue with a trigger follow through. That is, the finger continues to move the trigger rearward after the release. This is important because of the M1 trigger's design, and if not learned, has been known to cause doubling with some shooters, particularly those coming from a smallbore background who have been schooled to use a slow precise squeeze that stops as soon as release occurs. In the world of service rifle shooting, this is called "milking the trigger". I am not really sure where this term came from, and to tell you the truth, in a family magazine like this, I am about half afraid to ask. Anyway, it is believed by many top armorers that if the conditions are just right, and an effort has been made to lighten the trigger pull much below the 4½ pounds that the NRA and the National Board service rifle matches require, by stoning too much from the hammer's hooks and the trigger's sear, that the hammer can fail to lock at the rear of its travel. This can lead to the hammer following the bolt as it closes, resulting in a live round in the chamber, with the hammer down and uncocked.

Although this is embarrassing during a string of rapid, especially if you are winning, it is not as bad as having the hammer hesitate for a moment, and then release, causing a double firing of the rifle. For if it does this, some believe that the conditions leading up to a slam fire may not be out of the realm of possibility.

So the name of the game is not to try to get the trigger under 4½ pounds, and you would be very surprised at the number of people out there shooting really good scores with a trigger that is heavier than that, as many rifles will smooth out and slightly lighten with use.

Because of this, some armorers like to gather, swap, or start off with parts from rifles that have been dry fire cycled many times, as the phosphate parkerizing, which has a high friction resistance, is well worn off, and any high rubbing points easier to see.

If you want a top of the line trigger, the best way by far to get one is to contact a good gunsmith who has been schooled in making them.

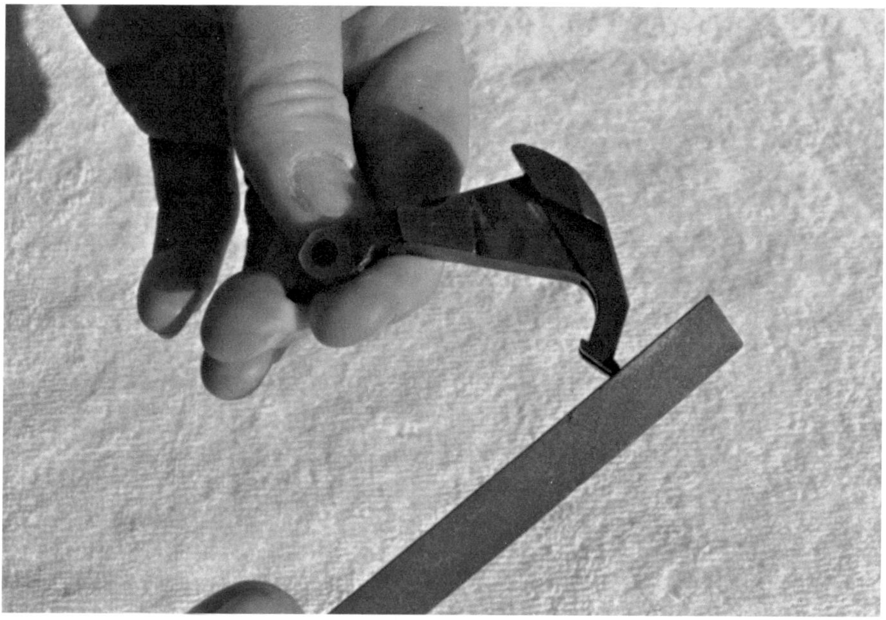

An armorer can lighten the M1's trigger by lightly honing the rear of the hammer's hooks. But their shape must not be changed.

Many of these good people advertise in these pages, and they are excellent. Do not hesitate to call them. They are worth it.

One trick that some do is to fit new precision fitting pins to the trigger and hammer, with some using a fit of .001 to .002 with a very high finish polish. Opinions differ on the pins (not surprisingly), with some making them out of the 4130 steel with a very light hardening. Others prefer different steels, or drill rod in its normal condition, feeling that a hardened pin tends to wear the inside of the hole in the hammer or trigger, and they would rather have the easily replaceable pins wear, as they should take far more rounds than the barrel, are easy to pull and relubricate, and it's a simple matter to check for wear with a micrometer.

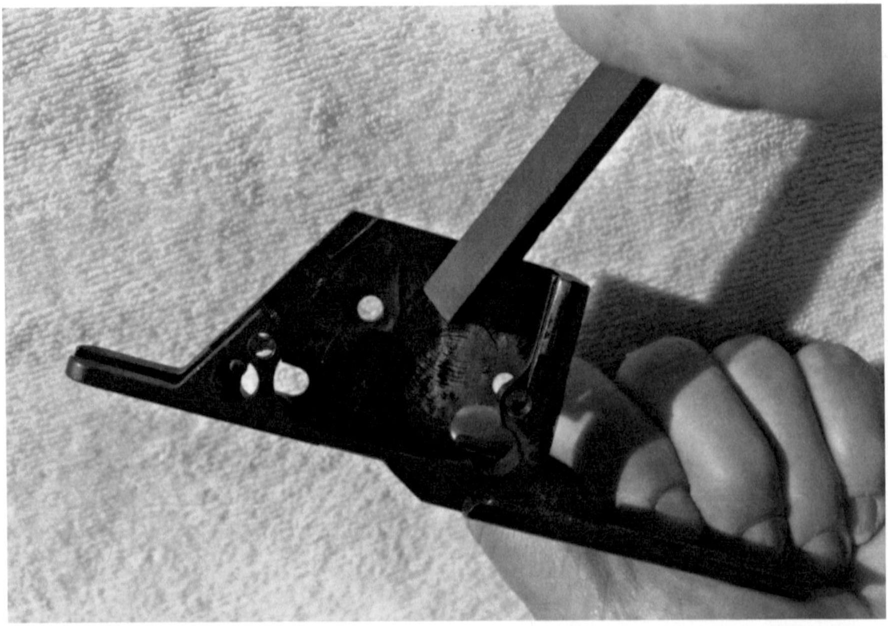

A fine India triangular honing stone is used to smooth the inside of the trigger housing.

If you decide to do your trigger yourself, you of course assume all risks. And you should seek a good gunsmith.

The armorer starts off by weighing the trigger pull while in the rifle, and logging this information. He then removes the trigger group, and while holding it in his left hand, uncocked, depresses the back of the trigger sear assembly to help relieve the spring pressure on the trigger's pin, and presses this out from the left with a punch. The hammer spring is strong, but can be contained with the hand if gripped firmly. He then removes the hammer pin and hammer. The trigger guard is moved back to the safety, turned sideways, and off the housing. Snap the safety to the "off" position, and remove by pushing outward at its top. The 8 round clip's ejector spring is removed by slipping a small punch or screwdriver in under its hair pin shape, through the small window on the left, and

Fully assembled, the M1's trigger is simple, safe, and reliable.

Taken apart, the M1's trigger looks more complicated than it really is.

lifting and snapping it out. All parts are then cleaned and inspected with a magnifying glass for cracks, wear, or other damage, and replaced if not up to specs.

Both pins are then polished, unless they are to be replaced. the polishing of most parts is done with 320 grit emery paper wrapped over small flat files, and with a fine triangular India honing stone, finished off with crocus cloth. Both pins go in from the right side.

The trigger housing has its flat inside walls that bear against the hammer and trigger polished. What the armorer is after is a flat, smooth polish where parts rub against each other. He is not trying to remove metal. It is very important to understand this, and most feel that it is better to just remove the phosphate parkerizing, and not the tool marks, as these serve to hold the lubrication for the parts. This is one of the rare cases where less is considered better, as the parts will wear and seat themselves with use. The next thing on the trigger housing is the sides of the rectangular hole that the trigger passes through. Some do the wall that the safety rides against, but some do not. Most triggers will rub one side of the housing when pulled.

A micrometer and machinist's hole gages are used to make precision fitting pins.

The hammer has its flat sides that rub against the trigger housing, around the hole where the pin goes through, polished with emery paper held flat and straight over a small file. Then the area on the flat surface of the left side, that the safety engages, is done. The horizontal groove at the rear of the hammer, that the nose of the hammer spring's guide rides in, is polished by wrapping the emery paper around a 9/64 drill's shaft and held by fingers (no motor drill).

This is about as far as most armorers will go with the standard GI hammer. Only in very rare individual cases will work be done on sear or disconnector hooks, and though some will polish the inside of the pin hole, many tend to use the micro roughness to hold its lubrication. But understand that these are opinions.

The hammer spring's guide has its nose (that mates with the groove in the back if the hammer) polished, along with its shaft. Also, the inside of the spring's round housing is polished, as are its sides by the trigger pin hole.

The sides of the trigger, where they pass through the rectangular hole in the trigger housing, are polished, and then the armorer decides what he wants to do with the trigger's sear. This is precision work, and usually done with magnifying glasses, or a jeweler's loop. If the sear is worked on, its edges will be kept clean and sharp, and its shape will not be changed. This is one part of an M1 rifle that is best left to those who are absolutely certain of what they are doing. If, however, you go boldly forth, and screw up, replacement triggers are easy to get at gun shows, or from parts dealers. Only the phosphate coating is removed if it is honed and polished.

At this point, all parts are cleaned, any rubbing against each other lubricated, and all reassembled. Some like light grease for lubrication, some swear by Teflon or moly bearing oils. Most did not oil the sear surfaces except to wipe them with a lightly oiled cleaning patch. This was to prevent rusting, rather than to lubricate. If a pin shows signs under a magnifying glass of rotating when cycled, it is lightly staked at one end. Again, we have opinions.

Now the trigger is reweighed. If slightly over, some will suggest that the shooter try it for a while then weigh again. Some will strip it and rework it. If, however, it is under 4½ pounds, they take a 1" x 10" strip of emery cloth of about 320 grit and place it between the back of the trigger and the disconnector. This is with the emery side against the

The simple act of snapping on the M1's safety prevents firing in three ways. Here a young student at Camp Perry practices locking his.

trigger, everything assembled, and the hammer uncocked. The emery strip is held taut by its ends so it only touches the back of the trigger and is slid back and forth for about 4 inches. They stop and weigh the trigger after every two strokes, until it will hold the weight.

Some armorers will aim for a pull of about 4⅝ to 4¾ lbs., rather than the red line of 4½, to avoid any chance of the rifle failing to pass the weight lifting test just before an important match. If it is heavier than they want it, they can adjust it by a very light honing, usually only a few strokes, applied to the rear of the hammer's hooks where the disconnector pushes against them to affect the trigger's release.

It is here that they stress that if this is done, the hook's edges must be kept clean and sharp, the angles must not be changed, and they must bear evenly, and with the utmost security against the disconnector. The trigger is then again checked for safety and weigh of release.

It is a simple matter to make an accurate trigger weight, using an empty can with a wire bail handle, hooked to a ¼" steel rod bent at 90° where it crosses the trigger. The weight must be lifted so that it is free of the rifle, and parallel to the barrel. The 90° where it crosses the trigger must be maintained the whole time. The rod and can are set on an accurate scale, and the can is filled with nuts and bolts until the desired weight is reached.

If not changed in design, the M1's trigger is among the safest in the world, and the simple act of snapping the safety on does three things. It grips and holds the hammer so that it cannot move. At the same time, it cams it totally out of engagement with the sear and blocks the forward movement of the sear, which of course stops the rearward movement of the trigger, as the two are connected.

Scientific tests have proven beyond a shadow of a doubt, that the M1's hammer time is substantially faster than that of a cave man's stone club swung overhead with two hands. Both have the redeeming quality of one main moving part, that is 100% reliable in ice, snow, mud, or 50° below zero temperatures. Both are slow, but effective.

These truths are borne out by the fact that man now populates this planet, and that a lot of men returned from World War II and Korea that would not have, had the M1 not had the reliability and accuracy that it has, while those who were up against it did not. It can, however, with proper follow through, and fine tuning by a good gunsmith, be made to give excellent results.

The search for accuracy never stops.

(Editor: The casual tinkerer is reminded that this article deals with a description of how a professional armorer fine-tunes an M1 trigger assembly. It is not intended as a do-it-yourself guide for the amateur.)

Low Cost Accurizing Tips for the M1

Part 10

Floating the Handguards and Gas Cylinder

(*Precision Shooting,* April 1995)

Like all modifications, all are not for everyone, and the ones that follow are probably more so than the others. However, since most GI M1 parts are readily available from gun shows or parts dealers, if one goofs them, it is really not too hard to get replacements. Also bear in mind that if you decide to do anything to your rifle, you assume all risks. Nuff said.

Not too long after the M1 made its appearance, it was discovered that many of them shot more accurately with their handguards removed. This is not surprising, considering all of the loose parts like the operating rod and its spring, etc., that hang on to its barrel and tend to dampen its harmonic vibrations differently from shot to shot. This gives a result rather similar to that of a guitar string that is plucked, then pressed (dampened) in several of the wrong places, at the wrong time. You cannot expect the results to be uniform, and accuracy is of course doing the exact same thing every time.

Some simplistic souls immediately decided to just plain take the handguards off, and leave them off. There were, however, a few problems associated with this. A few that quickly became evident: The rifle now looked kind of naked in the woods, and many of its working parts

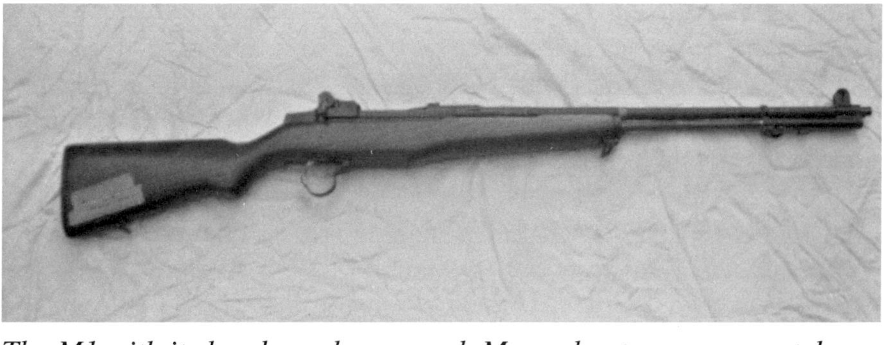

The M1 with its handguards removed. Many shoot more accurately this way, but not all of them.

were then exposed to dirt and the elements, while the barrel, when hot from shooting, tended to burn its shooter. Another little problem was that the mere act of taking them off transferred the rifle from the service rifle class to the match rifle class, where most rifles were bolt actions, with heavy barrels, and a tough act to follow with a stripped down M1. So the armorers had to come up with some other ideas which, being armorers, they promptly proceeded to do.

Some tried lighter weight wood, plus hollowing out the insides to the point where the handguards were virtually shells. It was very important after doing this not to let anyone except the shooter and armorer handle the rifle, lest the great secret be found out.

This somewhat classified information was not too hard to contain, for if the shooter owned the rifle, he or she had paid for this, and did not normally give tidbits of such information like this away. While, if the rifle was owned by the U.S. Government, the surest way to guarantee not getting your hands on it again was to open thy big mouth. Also the ire of one's superiors is not a thing to be cultivated if one is planning on being promoted in the military when on their rifle team.

When dealing with the M1, the government normally specified a certain grade of American walnut for the main stock unless they were really strapped, at which time they would accept this side of Balsa wood. The handguards, however, were a different matter, (read: Just About Anything Goes) and they were a bit more lenient on these. Hence, many M1's like this are seen from time to time today.

This gave the gunsmiths and armorers some leeway to experiment with, and naturally some did, trying other (lighter) woods, and staining them to look GI Eventually, the wood gimmick kind of fell out of favor as not really being worth all the hassle, and more emphasis was placed on getting better results with better barrels, and better shooters. The M14 and M1A missed out on a lot of all this by going to fiberglass, which is both thinner and lighter.

The U.S. Navy and others experimented with what is called free floating the handguards and gas cylinder. The term "free floating" is not really quite correct, as the parts still touch in places. Perhaps "semi-free floating" would be a better way to phrase it. Basically what was done was to keep as many parts from touching the barrel as possible.

This, along with some really good barrels, tended to shoot groups that were more round, tighter, and with fewer flyers at the outer edges of the group. Thus, when the group was well centered on the target, it produced some nifty scores.

But bear in mind that these modifications also have their drawbacks, so before you jump up and down, then run out and get it done like most of us do, consider that the front handguard and rear half of the gas cylinder are now somewhat unsupported, (kind of like my buddy's ex-wife) and any Garand so modified should never be picked up or handled by these parts ever again, for to do so risks breaking the front handguard just in front of the lower band, as this is what it will now be glued into.

Since it is perfectly normal for friends, relatives, sweethearts, and wives to pick up a service rifle by these parts, and you plan on remaining in the same state or country with them, it is considered very wise to label the rifle in some way so that this is known at once, before the rifle is ever picked up.

I have a strip of masking tape on each side of my M1's front handguard that has, "No Lift Here", printed on it boldly with red crayon. I have seen one with the word, "Never", printed in black on the top, and one at Perry that had, "Don't Touch This Darn Thing Here", in fine print

The front handguard can have its sheet metal strengthening liner removed to make it lighter. But doing so makes it weaker.

on tape on its sides.

To start off with these modifications, the lower band, where it fits on the barrel, must be a very snug fit. This is because it will now be the only support for the front handguard. To get this, some shim, some peen, and most fit a split, spring-type roll pin. Others fit a very tight-fitting solid pin. Joe Roberts Jr., who did an excellent article on these modifications in the July 1984 issue of the American Rifleman, suggested at that time using General Electric RTV-106 Adhesive from dealers in building materials, this would hold the high temperatures generated in rapid fire, which most epoxies will not. I personally prefer a tight mechanical fit in case I have to take it apart.

The rear end of the rear handguard has 1/16" to 3/32" filed or ground off of it, so that it cannot touch the front of the action. Some then glue the front of this handguard into its recess in the lower band. If this is done, both handguards and the lower band must be slid forward off the barrel, as an assembly, if they are ever removed.

Some do not glue here; instead they put grease in the grooves in the sides of the barrel that the handguard's clips snap into, feeling that the handguard will now center itself fore and aft, using the barrel's vibrations as it is fired. Still others do not grease, but slide the handguard as far forward as it will go, then stake the grooves so that the clips cannot slide to the rear. It seems that everyone has an opinion, and that of course is what improves the breed, for only through trial and error do we learn anything. I often think that the reason so many people get such good results with such varied ideas, is that barrels vibrate differently from each other, though looking the same.

Some then fit a 1/2" square piece of neoprene to the inside of the rear handguard, just thick enough to keep it off of the barrel, while others use a pad of epoxy bedding compound for this. Some place this pad right above where the clips snap into the barrel, while others place it just behind the clips, closer to the action's receiver ring. Now if you feel that this is getting confusing, and that there is a lot of room for experimenting here, you are right, for some omit this pad entirely, feeling that it does not do much towards damping the GI barrel's vibrations in a helpful way, as it is too far to the rear of the barrel, and tends to block the circulation of hot air between the barrel and handguard. So suit yourself!

The front handguard has the tabs on its sheet metal strengthening liner bent inwards and the liner slid out. This will make it lighter, and gives more room for the op-rod to vibrate in, and many liners sometimes show shiny spots where the rod has rubbed through the parkerizing finish. This is particularly true if heavy loads have been fired, causing the rod to bow, or if the rod was not correctly aligned. The liner is then left out.

Bear in mind that without this strengthening part, the handguard can no longer withstand any side blows such as it was originally designed

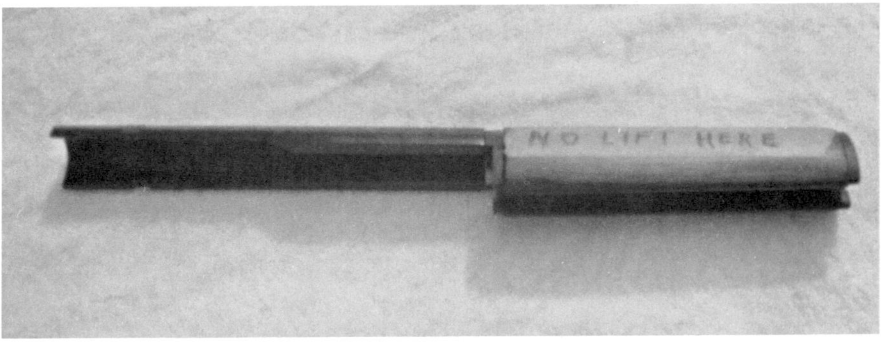

If the two handguards are glued together, and then semi-free floated, they become the rifle's weakest part, and it must not be handled by these parts.

The sheet metal cap at the front of the front handguard has the hole that the barrel passes through enlarged about .020.

for, although the metal caps at each of its ends tend to support it for normal handling. Glued into the lower band, it becomes the rifle's weakest part.

The sheet metal cap at the front of the front handguard now has the hole that the barrel passes through enlarged about .020", so that the barrel cannot touch it during its vibrations when fired.

Some enlarge this hole by boring, some by filing. But the slickest I saw was taking a drill with a #2 Morris taper on its shank, and pushing the tapered shank into the hole until it enlarged it just enough. But, the armorer that does this cautions that you have to be very careful not to go too far, as this will force the handguard's sides apart, and split it. There is however, and not too surprisingly, another way. More on this later.

At this point, the rear of the front handguard is glued into the lower band. Now some want this joint very firm, and will use strong epoxies backed up by #6 x 1/2" or so, wood screws inletted in from the other side of the band.

The good side of this is that the joint is strong. The bad side is that if it is dropped and broken, the wood will likely be damaged, making repairs at the range (which is where it most probably will occur) kind

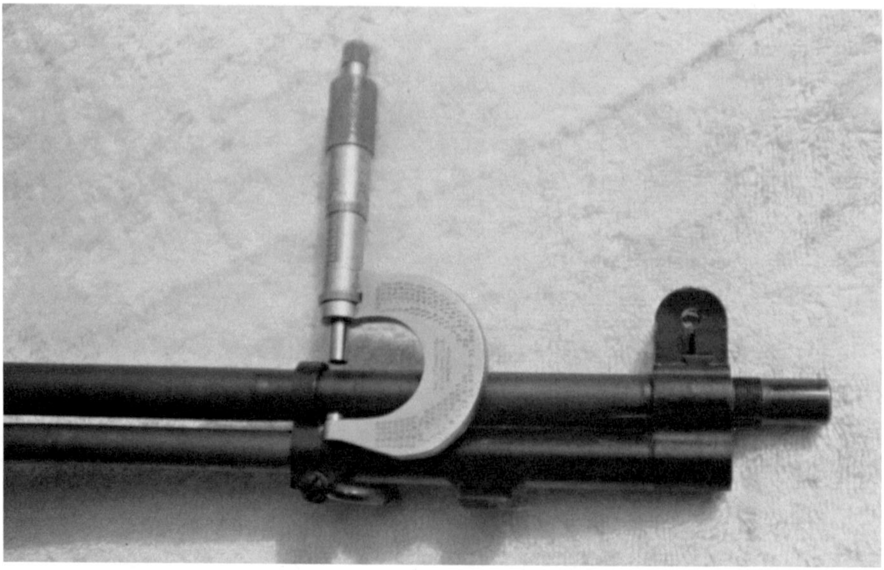

Some smiths turn the barrel down just under the gas cylinder's ring, and the front handguard's hole, to get their clearance.

of difficult.

Because of this, some prefer a frangible, that is, a breakable joint, in which the glue is allowed to fail before the wood does. But the glue used must be able to withstand the heat of rapid fire.

The gas cylinder must not contact the barrel except where it fits on its splines. Because of this, they must be a tight fit, and it must not contact the front handguard. To get this, about .020" to .030" is ground, filed, or machined off of the rear of the cylinder and its rear ring. If this clearance is not already there, and some have found that they can grind, or file metal off of the back of the gas cylinder lock ring, which allows the cylinder to be pulled forward when the gas cylinder nut is tightened. But if this is done, the gas port in the bottom of the barrel must still align with the port in the cylinder, and the op-rod's piston must remain in the cylinder when fully retracted.

The tab that protrudes about 3/8" from the bottom rear of the cylinder into the op-rod channel of the front handguard has its sides filed at about 30° to give it clearance, or just machined clear off.

The rear ring of the gas cylinder that goes around the barrel has its inside diameter enlarged by about .020" so that it clears the barrel all around. That is, if the barrel is .625" outside diameter (O.D.) at this point, the ring is machined to .645" inside diameter (I.D.). Usually this is reamed with an expandable reamer with a long pilot that locates in

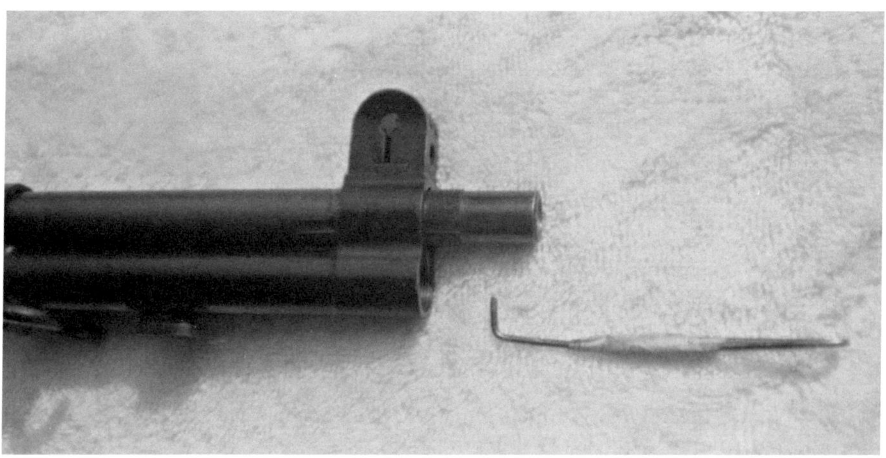

A coat hanger of about .075 thickness, bent 90°, can be used as a gage to check that the barrel and cylinder's gas port are still aligned if the cylinder is moved forward to get the clearance at its rear.

the three splines. But the operation can also be done by filing, or boring with a lathe or milling machine. If it is done this way, caution must be used as the web of metal between the bottom of the ring, and the top of the cylinder, (that is, the wall thickness of the cylinder), should not be thinned.

There is, however, another way. It ties in with the trick for the front handguard's hole mentioned earlier, and goes like this. Since the diameter of the barrel where the front of the handguard and the rear ring of the gas cylinder locate is about .624", while the diameter of the barrel just in front of this, between the gas cylinder's two rings, is about .600", some smiths just turn the small area of the barrel just under the rear ring and front handguard hole, down to where it blends in with the rest of the barrel, and do not enlarge the holes in the rear ring or front handguard.

This is pretty slick, and one smith takes it another step further for ease of handling. He takes a sheet of .008 shim stock 1⁷/₈" wide by 3³/₄" long, rolls it into a cylinder, snaps it over the barrel between the gas cylinder's two rings, then pushes it backwards until it is under the gas cylinder's rear ring, and the front of the handguard. This tends to take up the clearances, and give the parts support so that the rifle can be handled in an almost normal manner, especially while transporting it in its case to and from the range. He just slips it out before going on the

The M1 gas cylinder showing the rear completely machined off for clearance.

firing line, and replaces it when he has finished firing. This is somewhat similar to the old match target Springfield shooters slipping their Lyman 48 sight out of its quick release base before transporting the rifle.

All of the above operations should be performed by a good gunsmith or armorer, and all of the angles should be kept sharp and straight. But, and this is a really big but, because all of these cuts are to give clearance, their tolerances do not have to be as close as working on something like headspace. That is, the .010" clearance around the barrel does not have to be exactly .010". It can be more, or even slightly less. All that is required is that the parts do not touch while firing, and you may find the factory tolerance on all of these parts quite generous already.

One smith that I talked to had an opinion that I found very interesting. If the rifle is to have a really good barrel fitted for all out competition, then he carries out all of the above mods and dang the expense. If, however, you are going to stay with the GI barrel, and are very limited on what you can spend, then just turn the barrel down under the rear gas cylinder's ring and the barrel hole in the front handguard. Move the gas

The old M1 is still the choice of rifle for many entry level highpower shooters because of its low cost. This housewife loves hers.

cylinder ahead by filing off the back of the gas cylinder's lock ring, to get the clearance between the handguard and the rear of the gas cylinder, only if it is not already there, and only file on the sides of the gas cylinder's rear tab where it goes into the handguard's op-rod slot if it touches anywhere. He does not remove the front handguard's metal liner unless it touches the op-rod, which it shouldn't, or if you wish to lighten the parts. If, at a later date, you change your mind, you can wrap a piece of .010 shim stock around the barrel where it was turned, coat the inside of it with high temp epoxy, and clamp it in place with a small hose clamp until it cures.

Again, understand that these modifications were carried out in the past by top armorers trying to get every last bit of accuracy. They transform the M1 from a robust fighting machine into a very delicate target rifle, and it is questionable how much they will help a rifle with a standard GI barrel as they differ so much.

They can, however, be carried out by one with few tools. The rifle should not, however, give the appearance of having been smithed by an elderly, blind, Neanderthal woman, with a flint hatchet. Even though, the search for accuracy never stops.

Understanding the M1, M1A Sights

(*Precision Shooting,* January 1992)

There are many things that can be done with a standard M1 or M1A to make them more accurate. Some of these things are very simple, and can be done by anyone with ordinary skills. Other things require a good gunsmith. While still others sound complicated until you get into them and then discover that they are really not that tough at all.

What I will do here is to pass on to you bits of information that I got from various shooters and armorers and then you can decide how far you wish to pursue it.

Now the sights. When you click the elevation or windage knobs on your M1 or M1A rear sight, you expect to get one minute of angle changes, that is, about one inch of change per hundred yards.

But are you really getting that? Maybe. If you are after half minute changes you are asking even more. The reason is very simple. The sight parts on the Service Rifle are made with generous clearance tolerances. They have to be, to interchange with all the other rifles out there.

In spite of this, like the A2 sight on the AR-15/M16, it is the best service rifle sight in the world. Part of this is because the sight cover is actually a flat spring that takes up most of the play in one direction. There are a lot of micrometer target sights that don't have this feature. John Garand's sight was way ahead of its time. It still is. But there are some tricks we can do to make it work a little better for you.

If you are handy with a screwdriver and a pair of pliers, you can do this yourself. If not, you can take this information to someone who is, or a good gunsmith, and go from there.

Now bear with me for a little while because this is going to get a bit technical, but in the long run, you will understand your sight better. A machinist divides an inch into a thousand parts. They are written like this: .001, .002, .003, etc. He uses precision instruments to measure this. One of these instruments is called a dial indicator. It is a dial face like a clock, but has only one hand. A push rod on the back of the dial moves the hand. If you place one of these on your aperture and run it up and down and sideways, at the same time pushing the sight aperture from side to side, you may well find that it does not return to the same place each time. The higher up the sight is run, the worse it gets because it has less support there and it doesn't take much, for one minute on the M1 sight is .008 (8 thousandths of an inch). A 1/2 minute is .004. The average human hair is about .003, with redheads running about .004. So you see what we are up against. (This does not mean redheads are 1/2 minute people).

Naturally, some sights perform better than others in this test and most do very well because when shooting, you run them up or down and you don't push the aperture from side to side unless rotating the aperture for 1/2 minutes. Consider shooting some groups from a bench rest with your sight set in your normal way for 200 through to the longest range you shot at, say 1000 yards. You can do this at 100 or 200 yards.

Then, try the time honored trick of stretching a rubber band from the sight aperture to the elevation dial and repeat the test. This takes up the play in a downward and sideways direction and will work no matter how sloppy the sight is. If your groups are consistently tighter with the rubber band in place, you know you have learned something.

Armorers take the sight apart, anneal the base and squeeze it to a tight fit on the aperture, then using a fine grit lap the two together for a very close fit. It is well worth the effort. The base is the part of the sight with the two high guards on it to protect the aperture from bumps and thumps. It is hardened to resist wear and must be softened by heating and cooling slowly (annealing) or when squeezed, it will break. When the armorer reassembles it he gives it a fine coating of grease and it works very smoothly. If any wear develops later, he simply squeezes it tight and sometimes re-laps it. If you don't already have a hooded match aperture, this is the time to put one on.

The rubber band can be hooked over the windage or the elevation dial, but if you shoot a match with it on, it is wise to hook it over the elevation dial so you will have the windage knob free for quick changes of the wind.

If you don't have one of the manuals on the M1, by all means get one, even if you shoot an M1A. So you, or whoever does your work for you will be able to take your sight apart, clean and lubricate it. It is really quiet simple. Although the first time you pry the cover off, and it makes its loud snap that sounds like you broke it, you may have to have 2 or 3 glasses of warm milk before you go on. The whole sight is held together by the slotted nut in the windage knob.

If the sight is clean and well greased, it is very smooth when properly adjusted. Some of the armorers I talked to use white lubriplate grease, others use whatever they like. One uses a very thick automobile chassis lube in hot weather to dampen the side play mentioned before. They all say the same thing. Lubricate every part that rubs, slides, or turns against another part. Starting from where the bottom of the base slides on the rifle's receiver, right on up to the top, including the indentations the knobs locate in.

It is amazing the number of rifles out there that can use some adjustment. For many shooters, this is going to be old hash. But you can always pass it on to a new shooter or use it to light your fireplace. So bear with me some more.

The tension on both knobs is controlled by a small slotted nut in the center of the windage knob. It is click adjustable by a screwdriver. It should be adjusted just tight enough that if the aperture is run all the way to the top, then down a couple of clicks, you should not be able to push it down with your thumb.

You arrive at this setting by trial, one click at a time. All we want is for the aperture to stay up while we are shooting. If it is too tight, you will have to close the action or use a pipe wrench to turn the windage knob. Once you get it set, put a couple of drops of oil on the clicking parts if it has not been cleaned and lubricated.

Most shooters arrive at their zeros by counting the clicks up from the bottom. This is fine. But there is another way. John Garand put a screw in the center of the elevation dial to set this. Here is how it is done.

First, count the clicks up from the bottom to your 200 yard zero with national match ammunition or its reload equivalent. Write this down, don't try to remember it. With the sight set for this, grip the elevation knob firmly with a pair of pliers and loosen the screw in the center of the elevation knob. Now apply a sideways thumb pressure on the aperture so that it will not move and turn the elevation dial until the 200 yard mark lines up with the line stamped in the back of the receiver. (To find this line, hold the rifle like you were going to fire it. You will be looking right at the line.) Retighten the screw while still holding with the pliers. If you are normal, you will have to do it two or three times before you get it right because some of the parts will slip on you. But don't panic. No matter where the dial is, it will still take the same number of clicks up from the bottom to get your zero. But once you get it set right, you will find that the correct number of clicks for 200 will have the dial setting on 200, same for 300 and on up. Of course, it is not 100% perfect. You may have to come up or down a few clicks to get in the center of the bull at different ranges, but it is a whole lot better than nothing. I know one person who shoots the Palma Match this way and he always gets on at 800 in just a few shots. Same for 900 and 1000. It may not work for you, but it is what Garand had in mind when he built the rifle. A taped note on the stock can tell you the clicks needed.

Another trick top shooters do is to slowly file off the top of their front sight until their 200 yard zero is about 4 clicks up from the bottom instead of say, 10 or 50. This way, when they get back to the longer ranges, the sight is not so high up where it has less support and is subject to more lost motion. I see no reason why you could not carry this further and have a replacement sight filed way down for use only at the longer ranges. This would have the added benefit or bringing your cheek back down on the stock where it belongs. If you do try filing your sight, be sure to keep it square and straight. And remember, a hair is about 1/2 minute. If you change front sights, they have to be relocated in the same place, or your windage zero will change.

One last thing about the rear sight. It has some very fine threads in it. They are well protected from damage as long as the sight is together. If you take it apart to clean and lubricate it, treat the insides as precision parts because they are. Well cared for, they outlast their owner. Most of the M1's out there already have. Those people were probably very proud of the rifle you now hold. You should be too.

Of M1 Barrels and Violins

(*Precision Shooting,* March 1992)

It is not known by everybody, but if the operating rod on an M1 Garand is bent, and does not travel in a straight line when functioning, it will give big groups, and flyers aplenty. It is also not well known among highpower service rifle shooters that if you try to use very heavy bullets, and heavy charges of slow burning powder in an M1 Garand or M1A you will, in all probability, bend your operating rod. This is because powders much slower than IMR 4320 usually result in much higher pressures out where the gas is tapped off of the barrel and sent into the gas cylinder to push the rod back. This gives the end of the operating rod a hard smack instead of a long, gentle push, and thus bends the tubular rod.

When any rifle is fired, its barrel vibrates as the bullet passes up through it, similar to a whip, if one end of it is snapped up and down. This gets a bit technical but it is also similar to the vibrations of the strings of a beautiful violin when it is played. If you were to reach over, during a recital, and put your hand on the strings of the violin being played at the time, it would dampen the vibration cycles, and change the high pitched tones to a very dull sound. This would also probably get you thrown out of the theater... so perhaps you should simply accept our word for this.

Heavy barrels on bolt action target rifles are not as touchy as those on lightweight sporting rifles in this respect, but they ARE affected. That is why the vast majority of them are free floated, and not allowed to touch the stock's fore-end.

This brings us to the M1 Garand, and to some extent, the M14 and the M1A. The operating rod on the M1 is hung under the barrel from the gas cylinder at the front, to the action at the rear. It has to be long enough to tap the gas where it does because the powder used in the cartridge is already in service, so to speak.

But some smart armorers learned a long time ago that by careful bending and caressing you can make this situation work very well. The top of the rod has a lot of spring to it, but it can be bent the way that you want it to be if you take your time and go carefully.

The top armorers that I talked to at Camp Perry during the National

Matches told me that the rod should be fitted to the individual so that, with the bolt closed, the gas piston end of the rod should lay at the bottom, 6 o'clock position of the gas cylinder, with the rifle held horizontal. One can check this by resting the gas cylinder on a waist-high table, while the pistol grip of the rifle is held with one hand, and the fingers of the other hand reach under the front wood handguard and press upward on the operating rod, about a half inch behind the rear of the gas cylinder.

The rod should raise freely until it comes in contact with the inside of the gas cylinder at the 12 o'clock position. Expressed in simple terms... the dumb thing has to be able to be pushed up, and when released, it has to fall back down to where it was. Now it can't do this if it is full of fouling, because the piston is fitted to the cylinder bore by only a couple of thousandths. So it has to be clean in there, or our little test is worthless. This is test number one. Test number two requires one to remove the stock, trigger group, and recoil spring from the rifle. Then, when you raise the muzzle about 60°, the bolt, with the operating rod attached, should open all the way... simply because of gravity. When the muzzle is lowered about 60° below horizontal, the bolt and operating rod should close, all by themselves.

The breech end of the rod may be allowed to rub the bottom of the barrel slightly, and if it does, grease should be smeared on the bottom of the barrel where it rubs, to do away with any friction. The rod should not rub any other part of the rifle. If it does, clearance should be made; otherwise it's back to big groups and flyers.

The average Garand operating rod will be found to be rubbing the stock, one side of the gas cylinder, and just about anywhere else that it darn well feels like. But it can be adjusted by bending. One armorer clamps the rod to the top of his work bench, with a ruler beside it. He then pulls it so many inches, and it springs back so many inches. By determining how far he wants to bend the rod, he gets it where he wants it to be.

One other thing that goes along with all this is to always have the screw that holds the stacking swivel to the gas cylinder... tight. The swivel should be swung all the way forward, and the screw tightened. Never allow the swivel to swing freely, as it is believed to interfere with barrel vibrations. It should not be swung to the rear and tightened, as recoil will tend to move it to the front. It will probably stay to the rear if well tightened, but why tempt Murphy's Law... the one about "If something Can go wrong... it WILL go wrong"?

Not long ago a friend showed me some odd rolls of wallpaper that were on sale. They turned out to be most useful to a highpower shooter. The beauty of their use is that they can be quickly stapled to a target frame, with the color pattern facing away from you. You roll off what you want, and cut it off with a pocket knife. You then use a Magic Marker... be it red, black, or whatever color you might like, to make any sort of a bull's-eye that your little heart might desire. You can trace out a chicken for silhouette, or a 13-inch bull for high power iron sights, or a one-inch bull for varmint rifles. Shade it in with your marker, and blaze away. A roll of masking tape makes good pasters. I had heard of this economical way of doing things some time ago. The first time that I tried it... I really liked it!

DCM Clinic

(*Precision Shooting,* September 1992)

To my surprise I find that few people know what the DCM Clinic is. DCM stands for Director of Civilian Marksmanship programs. At the present point in time the Director is Colonel Cullinane. The clinic itself is a two-day school, held at Camp Perry (near Sandusky, Ohio) each year, sometime around the end of April. It is put on by The Ohio Rifle and Pistol Association, and the Ohio National Guard. The school is open to the general public, and costs approximately $30.00. Its purpose is to teach shooters how to compete in the National Match course of fire with the service rifle.

The school supplies almost everything. The issue rifle is the M1 Garand with M2 150 grain ball ammo. If the student finds this rifle a bit big, the Guard will instead issue an M16 that is modified for semi-automatic fire only. Each student has an instructor/coach (line officer) that stays right beside them on the firing line at all times. Safety is so firm and strict that it has to be encountered to be believed. All of the people running the school are volunteers; if it were not this way few could afford it.

So, how many students are to be expected at the Clinic? The April, 1992 class, which was bedeviled by wet, cold weather, nonetheless had 540 students attend. When one considers the poor weather handicap, that's a fairly impressive turnout. The students come to learn about

A very small portion of the over 500 students on the line at Perry.

offhand, sitting, and prone positions, rapid and slow fire, from 200 to 600 yards.

The students themselves are very interesting. It used to be that the average student was a male, somewhere between 20 and 30 years of age. Over recent years this has changed. We now have a mix of male and female, with many under 18 years. Some of these come from air-gun and smallbore programs put on by the Boy Scouts, American Legion, Future Farmers of America, 4-H, and on and on. This shows that these programs are working, in spite of the beating they are taking from the anti-gunners. The shooters of tomorrow are the ones we work with today.

We also get a lot of parent and youth pairs that come to get the safety training that is not taught in our public schools. And each year we get more and more police officers who quietly sign up and go through the school for highpower training. Consider that the average police officer today is born and raised in the city or suburbs... and yes, he may own a deer rifle, or perhaps a varmint rifle. But he has few places to go to really learn how to use it effectively. The average student today travels anywhere from 200 to 500 miles to attend the school. Many youth groups come in by bus. Some come from farther distances.

Although safety discipline is strict, it does not mean that you will not come away feeling that you had a good time. Being a volunteer is full of fun surprises. It is easy to "draw" a big, burly male student dressed in camos, who never misses a gun show... along with a teenage girl from a smallbore program, who might weigh 98 pounds dripping wet. Guess which one will have the recoil shy flinch? You're right, it will seldom be the girl. The flinch is cured by the instructor not letting the student know when the rifle is loaded during a string of dry firing when the line is firing prone, and everyone else is firing live. The student is made to concentrate on keeping a perfect sight picture before and after the rifle snaps, with the front sight in sharp vision. The instructor cycles the action while the student is ordered to look away. Somewhere along in the loading the instructor inserts the live round. With all the other students firing, it can be hard to detect. When the round is fired, it is almost always in the bull, and the student is so overjoyed that he cannot remember any recoil. This tends to build self control.

About all the student has to bring to the school is eye and ear protection, a mat or pad to lay on during prone firing, a spotting scope

Anyone who has ever been to Perry will remember the pits.

or binoculars, a shooting coat or maybe a GI field jacket, gloves, and rain gear in case it rains.

The school starts with a slide show and lecture instruction in the post theater at 7:30 a.m. Saturday morning. This will cover safety, range commands, duties and responsibilities of competitors, positions, aiming, zeroing of the rifle, trigger control, sight adjustments, effects of weather, and how to score and mark targets. This lasts two and a half hours, and includes a question and answer session.

The students are then squadded into two person teams with relays 1 and 2 reporting to their instructors on the firing line, while relays 3 and 4 go to the pits to pull and mark their targets for them. The instructor puts relay 1 on the line, with relay 2 scoring for them. If you came by yourself, you will be teamed with someone else who did the same.

The course of fire is the National match course, modified to the point where it is almost reversed. Over the years this has proved to be the best way to get new competitors started, and build confidence in themselves. The DCM Clinic course is as follows:

Sighters and 5 shots slow fire, prone.

10 shots standing to prone, in 70 seconds (rapid).

10 shots standing to sitting, in 60 seconds (rapid).

5 shots slow fire, standing (offhand).

It is not uncommon for an instructor to draw a student of any age, build, or sex that he may decide to start off prone, over a sandbag, until they gain confidence. Again, over the years, this has been proven to be the way to go. The M1 Garand can be intimidating, even for a smallbore student. A lot of times just removing the Garand and issuing a lightweight M16 is the answer. An amusing scene is the middle-aged housewife, coming off the firing line, raving (happily) about the M16 that she has just fired, and insisting that her husband will simply have to get her one. Those who have been competing in smallbore, and have already learned sight picture and trigger control can be real eye openers at 600 yards, slow fire prone. The clinic is open to all junior shooters 12 through 20 years, and all adult shooters who do not hold a classification in highpower.

A group of police officers watch as an instructor shows them how to close the bolt on an M1 Garand.

Housing for the weekend is up to the competitor, but most stay in the huts. These are rows of small houses, about 13' by 13', that hold 4 GI cots. You can get them for about four dollars, and the kids love them. You bring your own bedding and towels, and the showers are nearby. These huts were used to house POWs during World War II, and are where the average competitor stays during the National Matches for pistol, police pistol, smallbore and highpower. Indeed if you know anyone who has been near the top of their class in one of these categories, they have probably been here. Camp Perry is so rich in history that it probably could fill a good-sized book, and has done so. All four ranges are named for Medal of Honor winners.

On Sunday a two person team practice match, with awards of medals, is held. The course of fire is similar to Saturday's, and includes 20 shots slow fire prone from 600 yards. Again the instructor stays right with the student on the firing line. It's a bit of an adage at the Clinic that "the students may come and go, but the instructor never leaves the rifle". The bolt is always latched open, the safety is always on, and a chamber flag is always in the chamber. Everything that the student will do is preceded by one or more simulated run-throughs first. This gets the student settled in, and alerts the instructor of anything out of the ordinary to watch for.

The M16 issued is the old standard model, not the newer A-2. The sights have to be adjusted with a bullet's point, or a special tool. Because of this, and the wind's effect on the light bullet, the M16 is not normally used at 600 yards. The M1 is used instead. At this year's clinic (April, 1992) the 600 yard firing line was too wet to be used, so everything was fired at 200 yards. But you can use an M16 at 600 yards if you want to.

Without the help from the DCM, this program would be impossible. And the whole DCM operation is in great danger right now. Our lawmakers have once again decided to cut back on defense and this could be the last year for Camp Perry. If you feel that these valuable programs should receive further funding, write or call your representatives, and let them know your wishes. If you need more information, call the NRA. Remember, if you don't do it, it won't get done.

For more information on the Clinic, get in touch with:

O.R.P.A. Highpower-Barb Smith
PO Box 11490
Columbus, OH 43211

Mysterious Flyers from the M1 and M1A

(*Precision Shooting,* July 1993)

If you happen to watch a top service rifle shooter, or armorer checking an M14 or M1A just before an important match, you may see them very carefully raise and lower the muzzle of the rifle about 60°. They usually do this in a rather quiet place, and you get the impression that they are listening, and feeling for something.

The truth is, they are. What they are checking for is the feel and click of the well-oiled gas piston sliding smoothly, back and forth in its cylinder. For they know that if it sticks, or hesitates substantially, their groups can open up. That is, they can get flyers. We usually think of a flyer as being way outside of the group. But as any benchrester can tell you, with tears in his eyes, just one shot slightly out of group can cost you the match.

How this happens with the M1A/M14, is interesting. The operating rod that works the bolt hangs under the barrel in a neutral position. When the rifle is fired, it cycles the action in a smooth, though fast, timed manner. If the piston is sticking in the gas cylinder due to carbon or rust build-up, through lack of cleaning, it will not cycle the operating rod smoothly.

As the bullet goes up the bore, it passes the gas port hole, and the gas pressure behind it flows into the gas cylinder. If the piston sticks, the pressure needed to make it move will climb to a different level, and this can vary from one shot to another. Then, when the sticking is overcome, the piston slams into the operating rod. This disturbs the barrel vibrations, and they are no longer similar from shot to shot. Accuracy, of course, means doing everything the same. We now have a flyer coming.

The way to prevent this, of course, is to do what the smart shooters do: clean the gas cylinder and piston with a good powder solvent, wipe dry, and lightly oil, every time the rifle is cleaned after firing. Do not use abrasive cleaners.

Another way to invite flyers in both the M1 and M1A/M14 is to loosen or tighten the gas cylinder plug without supporting the gas cylinder with a separate wrench that is shaped to fit the cylinder. These wrenches are usually made out of aluminum or other soft metal to avoid marring the gun's finish. They are made to fit around both the barrel and

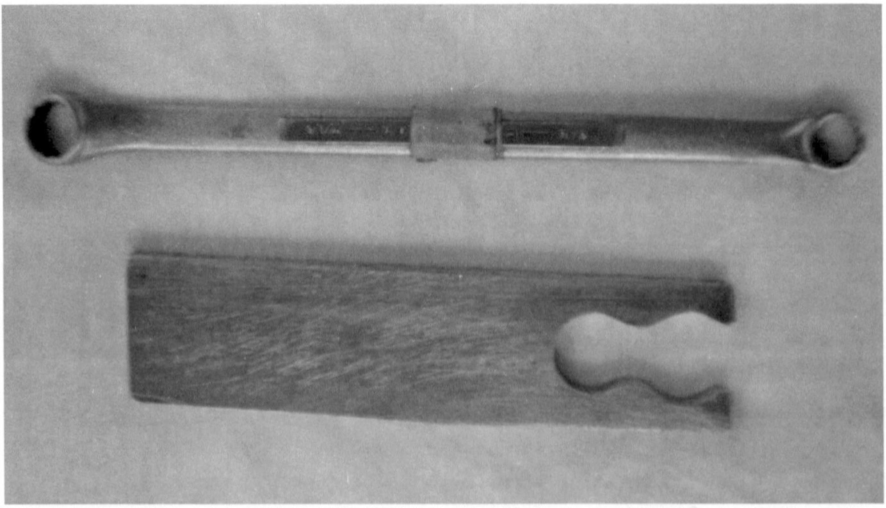

The hour glass wrench, and a 3/8 box end. These are the correct tools for removing the gas piston in the M1A/M14.

the cylinder, with a cut out that looks like an hour glass. I have heard them referred to as figure 8 wrenches, hour glass wrenches, and sometimes even gas cylinder holding wrenches. The gas cylinder on a match rifle should be a snug fit on the barrel's splines. Any unsupported twisting motion (torque) on it can cause looseness, or misalignment, which will cause it to push against the operating rod, and here we go again.

The military manual says to tighten this plug to 120 inch pounds. But the important thing is to keep the threads clean and well greased, and never, ever, over-tighten one, because you can usually twist things out of whack this way!

Most good service rifle smiths can supply you with a gas cylinder holding wrench. It is indeed money well spent.

The M1 Garand is equipped with a stacking swivel at the end of its gas cylinder. Its original purpose was so three or more rifles could be hooked together to form something similar to a tripod instead of laying the rifles on the ground.

If you let this swing loose on some rifles, it can also interfere with barrel vibrations and cause flyers. To be on the safe side, because not all M1s are affected by this, swing it to its farthest forward position and tighten it firmly. The reason it is swung forward instead of to the rear is

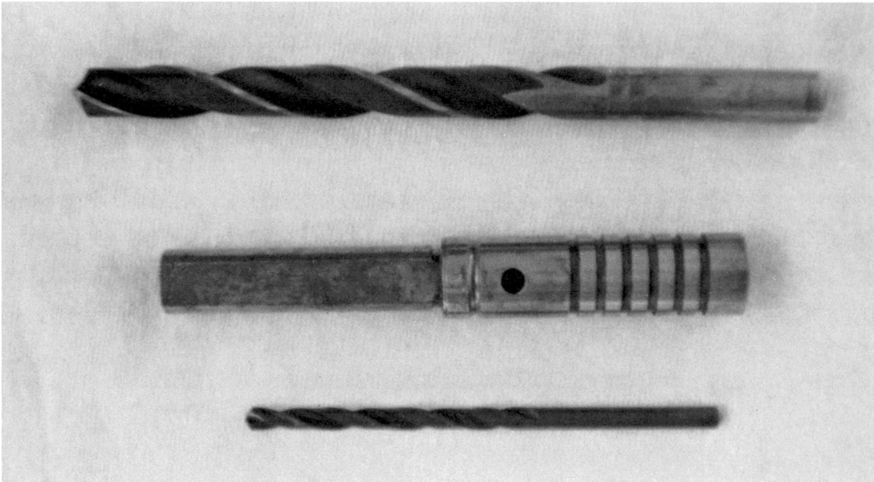

The M1 gas piston and the #26 and letter "P" drills used to scrape the carbon out of it.

because that is the direction recoil will take it if it comes loose. So you try to outsmart the dumb thing, and Murphy's Law, by taking up all the slack in the correct direction first.

You may see some rifles in matches with this swivel removed. The rules state that it must be on the rifle, but your friends probably won't complain unless you win.

The theory of how something as disconnected from accuracy as a loose swivel is worth some thought. It goes like this.

The rifle is fired, the bullet starts up the bore, the rifle begins to recoil rearward. The loose swivel that was hanging straight down by the effect of gravity, is swung violently forward in an arc coming to the end of its travel, it smacks up against the bottom of the gas cylinder, which is attached to the very end of the barrel, where it is subject to the most whip.

This would be like standing next to a benchrest shooter, and just as he fires, you smack the end of his barrel with a small hammer. (Of course then, it might be smart to take a fast, very fast, horse out of Dodge City).

But this makes it easy to understand how the slightest bit of friction from dirt, rust, or a partly tightened screw could change the swing of the swivel, and so have an effect on the accuracy of the M1 rifle. Very interesting, indeed.

The gas cylinder on the M1 should also be cleaned and oiled. Probably the best advice I ever heard on this, was to treat it like a smooth bore pistol. Clean it with a good chemical cleaner like Hoppes or Shooter's Choice, dry wipe it, and then oil it lightly.

It is surprising how easy it is to place your fingers against the rifle's operating rod when in the shooting position. It is even easier to lightly squeeze it with a shooting glove and not know it when concentrating on

This shooter is about to get a flyer. The fingers are squeezing the operating rod.

all the other things going on. This is a sure way to ask for a flyer, and sometimes the simple answer as to why two or more shooters get drastically different results with the same good service rifle.

What happens, of course, is that the squeezed rod pushes against the side of the gas cylinder which again dampens the barrel vibrations. Which is about the same as having the operating rod rub the stock after someone, somewhere, went to a lot of trouble to be sure that it did not do that.

New shooters can make this mistake, by not yet knowing how to hold the rifle properly but not too often, as they are usually terrified that their fingers are going to go into the clearance slot in the stock and be chopped off and eaten by the rifle and they will never ever see them again.

No, it is usually the middle of the pack shooter who is trying very hard to go up in his classification that goofs, because so many things are not going right. All at the same time.

This is particularly easy to do when getting into the sitting or prone position during the rapid fire stage of the national match course. His position is slightly off, he knows it is off, but pressed for time in the 60 or 70 second match, he tries to do his best, and wiggle his way through it.

All highpower competitors go through this. For some of us, it never ends, and this is where the shooter is most likely to slide his hand into the wrong position. Especially after a reload that went wrong on the first attempt.

But competition shooters being the type of people that they are, the best in the world, some very long lasting friendships have been formed, just by talking about what went wrong later. As a matter of fact, some southerners have been known to take pity on a few Yankees and actually help them, instead of telling them the truth, that they were just lousy shots. They probably thought the Good Lord would help them for this in their old age. Truly, shooting is an educational experience.

More Thoughts on Slam Fires

(*Precision Shooting,* October 1993)

The article in the April 1993 issue of Precision Shooting on Reloading for the M1, M1A, and M14 by Randolph Constantine Ph.D., was very interesting and very thought provoking, particularly the information on slam fires. Allow me to throw another log on the fire.

If you put an empty, primed case in many of these rifles and close the bolt, then pull the bolt handle back to where the bolt is about $^1/_4$ unlocked, and pull the trigger, some of them will fire. This is particularly true if the case shows resistance to entering the chamber, such as a reload not being sized enough, as it is then forced firmly back against the bolt face. This is also true of many bolt action rifles. This by itself seldom gives trouble, but combined with other things, it may.

Some service rifle shooters develop a good follow through as they trip the trigger during firing. Some do it the same from shot to shot during slow fire, but may not during a string of rapid when their mind is very busy. This can lead to a condition sometimes referred to as "milking" the trigger. That is, the trigger is squeezed so slightly, and then held there as it trips, that the disconnector fails to catch the hammer, and it follows the bolt down. This is usually more common with shooters who have spent a lot of time in smallbore before becoming involved with highpower than those who just start out with the service rifle.

Where this gets more involved is if the hammer is only partly caught by the disconnector or sear and is slightly delayed before releasing. This can result in a double fire, or, if the timing is just right, perhaps a slam fire. The difference being in how much time the bolt had to complete locking before it was struck by the hammer. Still, if the cartridge was sized enough to enter the chamber smoothly, the spring loaded ejector should hold the cartridge far enough away from the face of the bolt to where even partial travel of the firing pin should not be enough to set it off until locking is complete.

It is believed that this is one of the reasons that slam fires occur mostly with civilian shooters firing reloads rather than military shooters firing tons of new ammunition.

Another thing that can aggravate this condition is trying to rework the trigger to get a lighter pull. The NRA service rifle rules state that the pull may not be less than $4^{1}/_{2}$ lbs. and at most important matches this is checked. Still, there are matches where it is not, but bear in mind that attempting to get it much lower than that can lead to the above mentioned problems if sear or disconnector engagement is cut down very much. The hammer hooks should bear evenly and their hold and release must be reliable.

Still another thing to be aware of is the primer. As it comes from the factory in its box, the anvil is not seated against the priming pellet, and the anvil's legs protrude out from the cup. Seating the primer to its proper depth against the bottom of the cartridge case primer pocket pushes the anvil up to where it is now right next to the pellet, and the sharp smack from the firing pin coming from the other side can set it off.

With this in mind, it is quite possible that a high primer in a clean pocket may not fire as the bolt is closed, but just be seated deeper. Though it may not give good uniform ignition if the anvil is not brought close enough to the pellet. This may depend on how freely the cartridge headspaces, and how far the strong spring loaded ejector holds the case away from the bolt face. Sometimes all you get from this is a flyer.

But if a primer is seated as far as it can go and still high because it has residue under it, that's a different story, for here the bolt face itself may well fire it if it is closed hard enough, as the anvil may be against the pellet.

Then there is the cocked primer; that is one that is seated crooked to where one of the anvil's legs bottoms in the case's primer pocket, pushing part of the anvil up next to the pellet, while part of the primer's face protrudes about the cartridge case. This one is much harder to detect if you hurry the inspection, especially if trying to produce a large quantity of reloads in a short time. Again, it can be dangerous, or produce a shot outside the normal group.

The crooked primer is easier to produce than you might think. Not because the presses or primer seating devices we use are not accurate and square, because most of them are, but because the fired cartridge cases we use may well not be. If you look at the rim of a case fired in an auto loader under a powerful magnifying glass, you may find it bent and deformed from the speed of its extraction. If you place this case in a shell holder and start a primer in it and note which way the case neck is pointing, then lower the priming arm and rotate the case a quarter of a turn and then press up on it again, and do this until you get back where you started, you may find that the neck of the case describes an arc. Sometimes a really big one, and putting a dial indicator on it can lead to cardiac arrest. In spite of all of this, the dumb things often do quite well.

Another thing to remember is that a lot of surplus GI Brass may have been fired in a machine gun. These normally have big long chambers so that the gun will not stop firing just because the ammo belt has been dropped in the mud or stepped on and bent while running. Also, if you buy only used match brass, was it fired in a bolt rifle with a tight match chamber, or an auto loader with a chamber more sloppy than yours? Assuming that it is good brass to start with, are you sure the rims won't get bent the first time they are extracted out of your rifle?

A slightly bent rim, when placed in a shell holder of a reloading press, could allow the sizing die to push the shoulder back a few more thousandths, resulting in more stretching of the case when fired. It would not take more than a few applications of this before the case reached its limits of stretching and then you would have case head separation.

If you are getting bent rims on extraction, careful selection of powder and charge weight could change this. I have seen some hot 1000 yard loads that bent the rims to hell and gone, and some mild ones meant to control recoil in 200 rapid that looked so good you would have thought they were fired in a benchrest chamber with a mild load.

A GI case fired in a machine gun may really stretch on firing. The designer expects this, and allows for this one time stretch as the case may well be ejected and left where it fell in wartime. If, however, we acquire it and squeeze it back to fit a minimum match chamber, especially for practice so we can save our good brass for matches, we have to be very careful how many more times it is allowed to stretch, and how far. I have seen some cases used over and over with non max loads until the necks split. This is not the correct way to go, but it sure has been done. One guy I talked to coats cases with bent rims with automobile polish and then fires them. He feels that this will allow them to shift back against the bolt face and straighten themselves. He may not be nuts. Very interesting indeed. (Editor: But then again, he may be nuts. The practice MAY be hazardous, and we are not prepared to recommend it to the populace at large).

Fine Tuning M1, M1A Match Sights

(*Precision Shooting,* September 1994)

On the back of the base of the M1/M1A's rear sight, there is a vertical line that hopefully lines up with the center of the indexing lines on the rifle's receiver. These are the rifle's mechanical zero lines. The average target shooter tends to ignore these and just zero his rifle for windage on a target, then paint a white line down through them to get his course adjustment, and then finish up with a horizontal line painted across the windage knob and receiver, to get his finish adjustment.

Different colored paint lines can be used to mark the M1's sight for coarse and fine zero adjustments.

Often red fingernail polish is used for the simple reason that it is more readily available, along with its remover, from some handy place like K-Mart. The remover is nice to have if a new load is discovered that results in a windage zero change. All in all, this works surprisingly well. The only major drawback being that you are not reasonably centered in the middle of your rifle's windage travel, and you could run out of windage adjustment at long range on a very windy day.

Mr. Garand made a provision for this, and it works quite well. You again choose the load you want. Often this is the Government National Match load or its reload equivalent. It might be the load that you use the most at short range, or it might be a load that the rifle likes at long range.

Try to set up under the most "no wind" conditions you can find. It is usually considered best to start out on a bench at 100 yards, and later double check at 200, or the longest range you normally fire.

Set the rear sight on its true mechanical zero, that is, with the long vertical line on the rear of the sight's base running right through the long line in the center of the indexing lines on the receiver. Then fire a group. Some use three shot groups, and some like five. The allen screw in the back of the front sight is then loosened, and the front sight shifted in the opposite direction that you want the bullet to move, a small amount at a time, until the rifle prints where it is held. The allen screw being snugged before each firing, and final tightened when zero is achieved.

Many shooters just use fingernail polish to mark their sight's zero. A bottle of remover is handy if a better load is found.

Bear in mind, that these movements of the front sight in its dovetail slot are usually very small, as it only takes about .008 of an inch to move one MOA on most rifles. If you are handy with a depth micrometer, you can figure this closely, but for most of us, the naked eye, or a simple magnifying glass works well. Probably the easiest to work with is a six inch scale (preferably a machinist's) and magnifying glass to measure the sight's movement from one side of its base.

Once zero is achieved, the paint lines can again be applied for quick reference, and any load used that requires a few clicks of correction can have this information taped to the side of the stock with masking tape, or logged in the score book (where it will be left behind with other important forgotten items), or taped to the body of the spotting scope.

Most match sights feature the rotating hood aperture that has the eyepiece hole offset about .004 so that half minute of an angle corrections can be had in elevation. The hood usually has a small "V" notch cut in the top of the hood that faces the shooter's eye. In theory he will always know if it is rotated to the up or down position by checking this with the rifle lowered from his shoulder far enough that his eye can focus on it clearly, or reach up and feel for the position of the notch with his fingernail if he is firmly in his sling position. (Providing of course, that he has not chewed these off in anticipation of the coming match.)

It is a slick idea to put a dab of white paint in the little "V" notch. Your eye will be too close to focus on it while firing, but you can see the white dot in the blur and know when it is in the up or down position. Sometimes a small piece of masking tape works well for some shooters. It is surprising how often a position, or even a match, has been lost by the shooter failing to remember the position of the hood from the previous shot fired, when the mind is loaded down with all of the other things it must calculate.

By the way, a small magnifying glass is very useful for looking at the rear sight's indexing lines on the back of the receiver. These small lines are about 1/32 of an inch apart (.032). So if your rifle is normally fired now at, say, two lines off center, that is the amount that the front sight has to be moved to center it mechanically.

Checking an M1 match sight for repeatability with a machinist's dial indicator requires some thought if you are not to feed yourself some false information, and you should always think twice before you run off and accuse your gunsmith of selling you a loose sight. The problem

A one inch travel dial indicator is excellent for checking the true movements of the M1/M1A's rear sight. If the windage reading is taken off the sight's base, the radius button on the indicator's stem can be left on.

starts with the fact that the sight's hood is round (about .460 diameter). So you are working with your indicator's stem against a .230 radius (half of .460). If you use a Starrett general purpose indicator that only has 7/32" of travel, you have to be very careful that you do not run out of travel without realizing it. Also, I have found that I can get a more accurate reading if I do not use any of the buttons of different radius on the end of the stem of the indicator. This is because the roundness of the hood and the roundness of the buttons tend to ride over each other as the sight is elevated, as it does this in an arc. However, if you work with small portions of the sight's travel and keep everything square and straight, you can do very well. I have found that the one inch travel Peacock indicator that I got from Otto, of O.K. Weber, who advertises in these pages, is excellent for taking readings of both elevations and windage.

Another problem that crops up from time to time with some ball locating 1/2 MOA sights is that sometimes "V" notches on the inside of the windage knob are not always of the same depth. This is a tough dimension to hold during manufacturing, as a 1/2 MOA on the M1 is about .004, so you are dealing with .004, plus or minus zip. Naturally, if the dimension is over, or under this, you will get more, or less than a 1/2 MOA per click of the windage knob.

The good news is that if you find one or more notches that do not

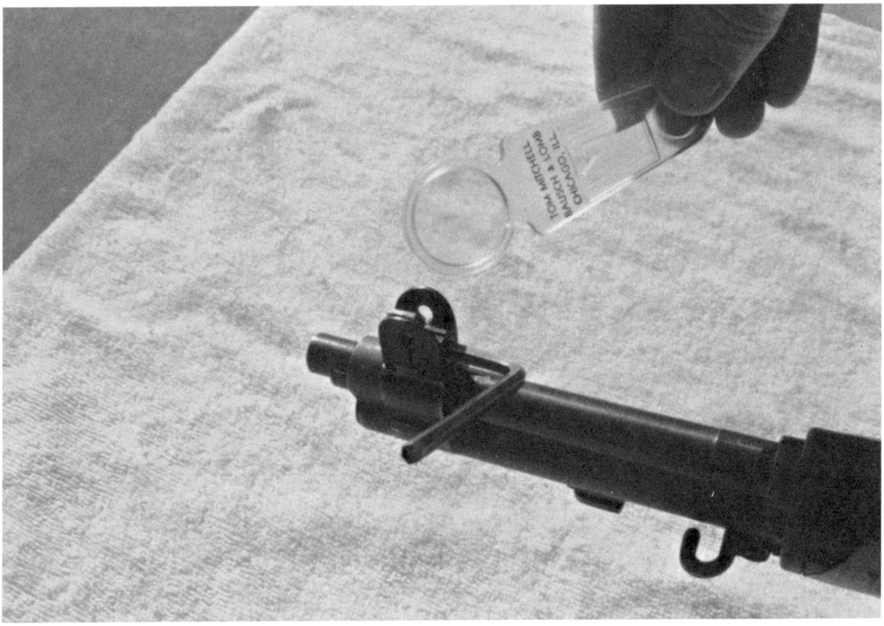

A magnifying glass and allen wrench are used to move the M1's front sight to zero it with the rear sight's mechanical zero marks.

match the other while doing a dial indicator travel check, you can mark their location on the side of the windage knob with a red crayon, and if they locate on a 1/8" hardened ball, which most do, start with the shallowest, and using a triangular shaped honing stone, very carefully, and very slowly, bring all of the offending notches to a uniform depth with the others, checking often with a micrometer as you go. The second piece of good news is that if you take .001 or .002 too much out of one of them, it is easy to bring the rest even with it.

The checking with the micrometer is interesting because if the windage knob's "V" notches snap over a 1/8" hardened ball, the smooth shank of a 1/8" is dropped into the "V" notch to simulate the ball, and

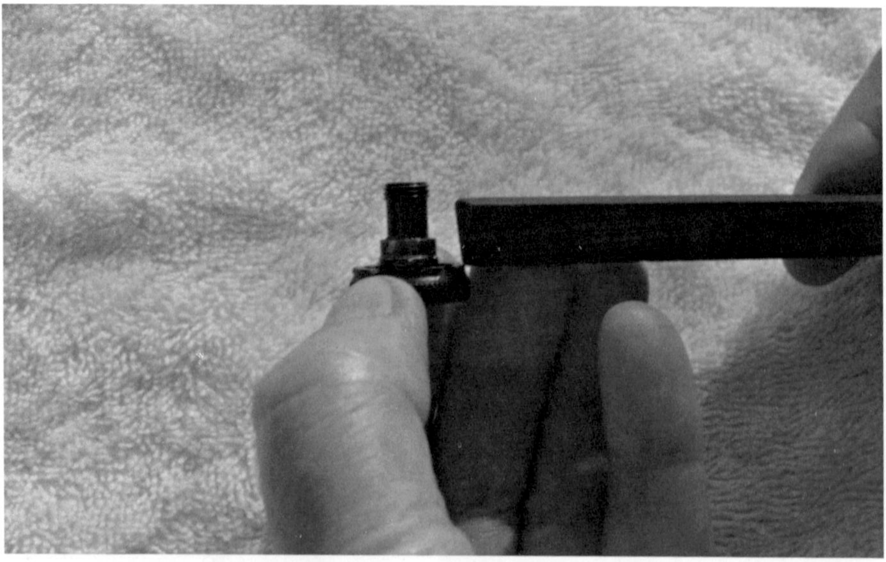

The notches on the inside of a 1/2 MOA sight's windage knob can be uniformed by light honing with a fine triangular stone.

the micrometer reading taken over both the knob's face and the side of the drill's shank. This is much more handy than trying to put another ball in the notch, and expecting it to stay there while you try to mike it. It is usually best to give each notch a number, 1 through 8, then count the number of full or half strokes that it takes with the honing stone on each one to reach the desired dimension. This is somewhat like counting the strokes of a cleaning rod while cleaning a very precision barrel, and is much easier than it sounds.

Once the notches are closed, the windage knob is put back on the sight, and the travel again checked with the dial indicator. For this is the only way that you know for sure how much the sight is moving per click. It is then re-lubricated, and the sight's spring tension is reset.

Most sights tend to get more wobbly, the higher up the elevation is set, because they have less support up there, and the M1 is no exception. So another simple trick is to file the top of the front sight down until the 200 yard zero is only 4 to 6 clicks up from the bottom. This of course reduces the rear sight's height at long range, and it is not uncommon to find long range shooters that have front sights so low that they are fitted and used for long range only, as they do not have enough travel to be used at close ranges.

Some shooters file the top of their front sight down until their 200 yard zero is only 4 to 8 clicks up from the bottom. This reduces the sight's height and wobble when set for long ranges. The top must be kept square.

Shooting a service rifle with its post front sight accurately requires that you see that front sight sharp and clear. If you are like most of us, and age has given you out-of-round eyeballs, do not hesitate to get corrective glasses that will allow you to see the sight sharply, for only in this way will you ever know how many points you have given away.

By the same token, if your eyes are so bad that you cannot see iron sights no matter what, just put a good scope on your rifle, go to your match and announce that you are there to shoot for your own scores only, and are not there to take away their gold. But do not be surprised if, in the following weeks you are joined by others with the same problem, and you find yourselves in friendly competition with each other on a very low scale.

I once took third place in a match like this, and the prize was half of a cup of coffee. But the people I was with made it feel like it was the finest trophy in the world. I went back two weeks later and tried very hard for first place, which was a full cup with cream and sugar. But I never could beat that old geezer, and how he loved to slowly pour out that half cup, was a scene to behold.

There is another thing that shooters do not, as a rule, pay enough

A 1/8" drill's shank and a micrometer are used to check the uniformity of a 1/2 MOA sight's windage knob's notches. This is easier than trying to use the micrometer and a 1/8" ball bearing.

attention to, and that is the placement of the head, and hence the shooting eye, in the same exact place for shot to shot repeatability.

On a shotgun, there normally is no rear sight, and the shooter's eye becomes the sight. So early on, top shooters who were "after" every scoring point to be had, learned that if they did not want any windage or elevation change with the center of their pattern, they had to put their head and eye in the same place each time they fired. Over the years, a lot of special and off set stocks were made to do this. They also worked with bending barrels.

With rifles, it is slightly different, but not a whole lot. With a telescope, the eye can be shifted up or down, and from side to side, and if the parallax is adjusted to zero for the range being fired, you are usually O.K. But iron sights are more critical, and the eye must be in the same place for each shot, or error will creep in. The eye should of course be centered with the center line of the sights, but this is, as a rule, not as critical as having it in the same place for each shot. This is not to say that your eye does not have to be centered when using a scope, because it does, but a really good one's (read: expensive) adjustments make it far less critical.

The bolt gun shooter goes to great lengths with his or her sight and stock design, and adjustments, to get as close as possible to not losing any scoring point to this problem. Most of these are capable of being changed for the different positions such as offhand, sitting or prone, along with adjusting to shooters of different size and shape.

The service rifle competitor, however, lives with a different set of rules, as there are only certain things that they are allowed to change, and where the problem really comes to be noticed is at the longer ranges, of 600 to 1000 yards.

Here, the rear sight is cranked up really high, and many shooters of normal stature find that they can no longer get the same cheek weld, spot weld, or head position that they want, and need.

The usual remedy is to slide the cheek up higher and higher on the stock, until in some cases, the chin is about on top of it. I know some people who are forced to shoot with their mouth held open a certain distance, in order to get the sight picture they want at 1000. In these cases, it is very difficult to get the eye back into the exact same position each time a shot is fired. I have experimented with this, and the results were not too bad, but you always had the nagging feeling that a bug was going to fly into your mouth and do slow rolls and loops while you were trying to get the shot off.

Another nutty friend of mine experimented with a prize fighter's and football player's mouthpiece in place so that he could clamp down on it and so have a position base stop. This worked pretty well, so we next tried kid's toy rubber balls of different diameters for different ranges, and got pretty good result, except that he kinda slobbered a little bit. We finally decided that what little advantage we were achieving was not going to be worth the social shock it was going to cause if we came out of the closet and went public with our secret weapon at a registered NRA match. However, if I were a sniper out in the boonies, I would not rule out more experimenting along these lines.

I still have the 600 and 1000 yard balls setting on my work bench by the drill press, and there are times when my mind returns to them. This is particularly so if I am working on a sight's thread, or trying to get something perfectly square in the lathe, and my wife of some 31 years, whom I do love very much, comes up behind me and says "Do you

We tried a kid's toy rubber balls in our mouths to hold our chin in the same exact place so our eye would line up perfectly with the sight line. You're right, it was nuts. But do not rule something like this out until you have experimented with it.

know what time it is? You know the high school reunion dinner starts in one hour. I'm not going if we're late, cause you know how everyone stares at you when you come in late. Sometimes I wonder why I ever married you in the first place.... Everyone said not to."

Yes, it is times like this when I look longingly at those two firm rubber balls, and my mind tends to lean towards the 1000 yard one, which is the bigger of the two. But then, you remember all of the good things, the shared hunting trips, and the target matches, and you just shrug and go get changed for the dinner, and put your dumb thoughts behind you. You would have to have a lot of extra balls for the judge and deputies anyways. *(Editor: I will **not** insert a quip here, no matter the temptation....)*

There are of course, other ways to deal with this problem. (The rifle's; not the wife's). One would be to try get the rules to allow a lace on leather cheek piece like they used on the M1 Garand "C" and "D" sniper models, and another that is in use, and works reasonably well, is to continue filing the front sight down as far as practical as this not only lowers the rear sight in its base as mentioned earlier to reduce its lost motion, but also will help get your cheek weld and eye back down where they belong. As a matter of fact, I know of one rifle that carries two filed down sights and the allen wrench needed to install them in the stock's tool roll under the buttplate, while on the stock's side there is a taped note that says, sight #1, 850 is now 1000, and 300 is now 600. Meaning that by fitting the filed down sight #1, the shooter sets it on 850 to get 1000 yards and sets it on 300 to get his zero for 600 yards. When front sight #2 is fitted, the dial is back to normal and set for 200, 300, and 600 to get those ranges for the National Match course of fire.

The problem being that you can only remove a little over 1/8 of an inch from the standard or Match front sights before the post is about gone, while the rear moves through about 1/2" of travel to get from about 200 yards up to 1000 yards. There has been consideration given to fitting a set screw with a smaller head diameter to the front sight, so that you could machine the platform that the post sits on, and then the post even lower.

No discussion on post front sights would be complete without a little bit on blackening them, so bear with me, maybe I can shed some light on this subject.

A season or so ago, my wife Mary and I were hunting woodchuck on a friend's farm. The farmer had an old aluminum bass boat with a flat bottom and his three kids had hitched their gentle old plow horse to it and let it tow them around the mowed hay fields. It was great fun for all concerned. When they finished, the farmer leaned the boat once again up against the side of the barn, with its bow pointed upwards, and its bottom facing us where we sat on a hillside quite a ways away. The bottom of the boat had become polished until it shone like a mirror.

As the sun went down to one side of us, it made the right-hand side of the bottom of the boat slowly get brighter and brighter until, with the naked eye, it was almost impossible to see the full width of the bottom of the boat. Had we gone over and painted it flat black, we could easily have seen the full width of the bottom again instead of only the right half.

The same thing happens to a post front sight. (Yep, they do get drug through the fields by a horse.) If one side of it, especially at the top, gets polished from rubbing, and the light comes from the right angle, you will normally only see the part that is polished, depending on the stretch, and angle of the light, which if from the sun, will be changing as the sun moves, and any clouds that get between it and the sight.

There has been consideration given to fitting a set screw with a smaller head diameter to the M14 front sight so that it could be machined lower.

If you can only see one side of the sight, you will in effect, make a windage adjustment to that side with your naked eye, because you can not see the other half, and you tend to center what you do see. If you do not stop this condition, you will, without realizing it, be making windage, and also elevation adjustments, however so ever slight, every time the light changes.

Black powder shooters learned this very early on, and also found that one of the best things to use to combat this was black soot from a fire. The only real drawback to this was that it rubbed off easier than it could be put on. But it was, and still is, a superb light absorber. However, there was another way. (Boy do I like that phrase, it sounds like to two semi drivers talking.)

When I was a very young kid, my grandfather, who built Match target muzzle loaders in Pennsylvania, used to take me for walks through the wooded hills. His rifles were all too big and heavy for me to hold up, so he would rest them over a log, or anything handy, and let me blaze away. He liked German silver for his front sights, but he taught me to always "whet" (sharpen) your sight if needed, which was often,

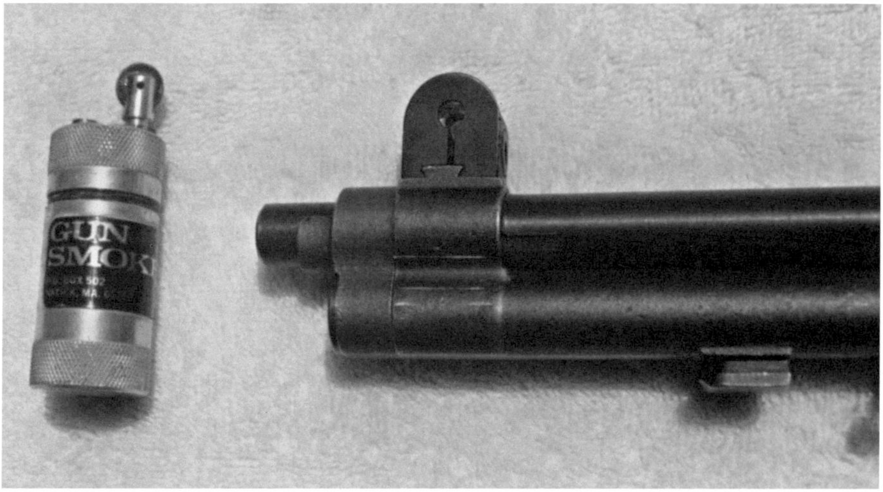

One of the best ways to blacken a front sight is to smoke it with a carbide smoker.

because it did not stay "whetted" very long with normal use. If the light was very poor, such as at morning, sundown, or cloudy, the sight was rubbed about one or two passes with your bare, dry thumb to make it light enough to see. But to whet it, you whet your thumb with your tongue, touched it to the ground, or any place else that you could get some dirt or blackening, then passed this over the front sight. It dried to a dull shine dampener in seconds, and while it was doing this, you wiped the mud off of your thumb on your clothes, so you would not get it on the rest of the rifle. The approved method was to wipe it on the side of your pants, about where your arm hung naturally, as any movement to wipe it on the front of your shirt could spook dear or turkeys or give away your position to the enemy. By the same token, if you did not wish to catch hell from Grandma for "Whetin' your sights on your pants again," you wiped it on the seat of your pants, where it could not be detected as this area would be dirty from sitting-in-wait in the woods anyway. But if you wanted the neighborhood kids to know that you had been out shootin', you wiped it on your pants, and let Grandma rant and rave. She gave most of the ranting to Grandpa anyhow.

Over the years I have heard the term "wetting" your wights, and it is true that you have to wet your thumb to get mud from dry dirt, but just as often, I have heard it pronounced "whetted".

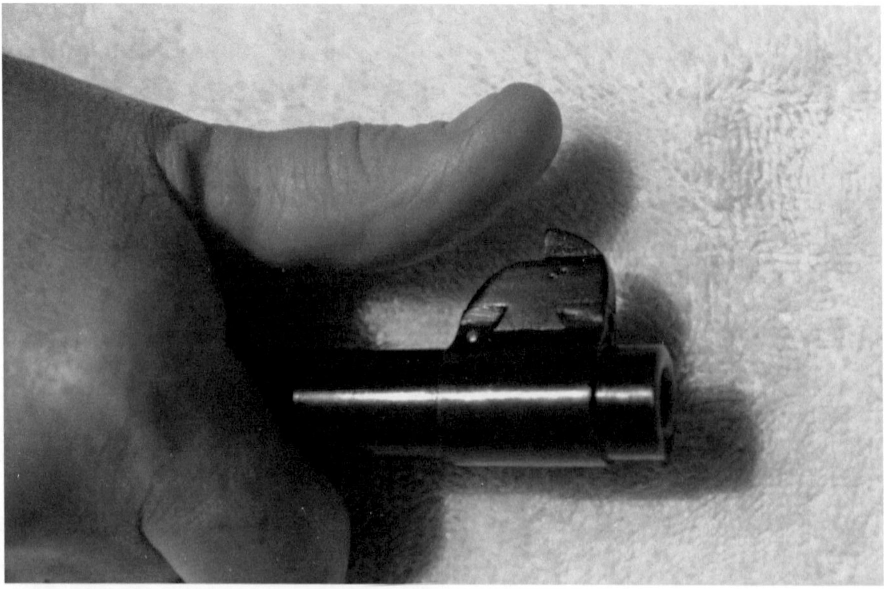

Contrary to movie information, wetting the '03 Springfield front sight with your thumb will not darken it, unless mud is added.

To add to the confusion, we had the movie "Sergeant York"... in which one of my favorite old time actors, Gary Cooper, portrayed the story of Alvin York, who became probably the most noted hero of the first World War by picking off a very large number of German soldiers with his service rifle, silencing several machine guns, and saving his unit as it was pinned down.

In this very good movie, several scenes are set up in which Cooper wets his thumb, then rubs his front sight with it. When questioned about this by his sergeant in once scene, he makes the statement, "It takes the haze off my sights." A polished post front sight will indeed appear to be hazy to the naked eye in bright light, and if bright enough you cannot bring it into focus as a sharp distinct object, which you must do if you intend to do any serious shooting, and adding plain water to it from your tongue will only make it worse. If, however, you touch the thumb to the ground only momentarily, and let it at once coat itself with dirt, you have more than enough mud for a sight. Heck, you got enough left over for your pants.

When you understand how movies are made, in which different areas have to be leased or rented with all of the hassle that goes just with this part, then they may shoot only one scene here, and can't come back for another month because one of the leading support actors is sick, or they cannot get the equipment to put in the background because someone else is using it, or a million other things, you begin to understand that they may postpone a scene you consider important, or move its location several times.

Under these conditions, the information given on how a sight has its haze taken off of it could well have been given months ago, and critical parts forgotten, or missed. By the time screening has revealed a problem, it may have been repeated until it would take thousands of dollars and re-opening of union agreements to correct it, and the people putting up the money felt that it was just not worth it. Whatever happened, it appears that the movie people did not get it quite right. But then, you only have to watch a little of what's on television to figure that out.

The search for accuracy never stops.

Slams and Jams with the M1 Garand

Slam fires, or the rifle firing before the bolt is completely locked closed, have been known to occur with the M1 Garand. Because of these incidents, more work and research is being done to find out just why this is happening. For it is a curious thing that the incidence of slam fires was almost unheard of when this fine old rifle was being used by the Military only. Indeed, it was only after large numbers of these rifles began to appear in civilian hands that the problem really surfaced. So the natural question follows, "What are we doing wrong?".

The answer is quite a few things really. For instance, we know that if a round is fed from a magazine, the resistance of this is high enough to slow the bolt's closing so that it just about never happens here. We also know that with a well maintained rifle using standard GI ammunition, we have no printed, documented reports of this occurring. So now, the investigation gets into handloading, modifying the rifle, worn parts, and the build up of clearance tolerances in parts that are not supposed to cause problems, but indeed can.

One of the first things learned was never to put a handloaded cartridge into an M1's chamber with an empty magazine, then trip the bolt and let it fly closed. For if you have a high primer caused by failing to seat it deep enough, or worse yet, the primer pocket was not cleaned sufficiently and has firm residue under it to keep it from bottoming in the case's primer pocket, it could indeed fire. If this is teamed with a dirty chamber, or a case that was not resized correctly, once again, it can fire. Most of the smiths and armorers that we talked to felt that if the bolt was let go from the half open position, the danger was lessened considerably. And if the handload was made to the same specs as GI ammo, the danger was just about nil.

The investigation then turned to modifying the rifle, and here things got interesting. The NRA and most other sporting organizations forbid lightening the M1's trigger pull below 4$^1/_2$ lbs. And for good reasons. Most think that if the contact between the sear and the hammer gets too small, the hammer will just fail to stay cocked. This will follow the closing bolt with no problem other than you now have a live round in the chamber with the decocked hammer resting against the firing pin. And that is usually what happens. However, if the lightly held hammer is held cocked until just the right instant, and then jars off the sear, before the bolt is fully locked closed, nasty things can happen.

The design of the rifle is such that the hammer is not supposed to be able to drive the firing pin all the way forward until the bolt is fully locked closed, and everything is then in alignment. This is where too loose, or accumulated tolerances come into the act. For if the bolt is loose enough in its guide rails, and if the firing pin is loose enough in the bolt, and/or the hammer has enough side clearance, the parts can sometimes shift enough so that the firing pin does line up with the hammer.

To test this, a solid gas cylinder nut was drilled and threaded for a 1/4-inch bolt. By screwing the bolt inwards until it pushed on the gas piston

A solid gas cylinder nut was threaded and a screw was fitted. This was used to keep the bolt from completely closing.

on the end of the operating rod, it was possible to stop the bolt's closing anywhere from 1/4 locked, to fully locked closed.

Empty, primed cases were then placed in the chamber and the bolt was tripped from the full open position. This test was tried on four different rifles, all of which had 4 1/2 to 5lb. trigger pulls. No primers fired, which was exactly what was expected, as my inquisitive friends proceeded with their test.

Empty, primed cases fired in a rifle with a known large chamber were then tried so that their failing to chamber fully would allow the extractor to snap over the case's rim instead of pushing the case ahead of it. The trigger was then pulled, and in three of the rifles, the primers could be fired with the bolt 7/8 of the way closed.

By screwing the stop screw into where the bolt ended up 3/4 of the way closed, and the extractor snapped over even full sized cases, it was often possible to get at least one rifle to fire the primer by pulling the trigger in one to ten tries. Sometimes it took longer.

These were all rifles that had been gauged, and passed as fit. So what does all of this mean? It means that any time a bolt is not fully closed, do not pull the trigger. This of course is normal safety procedure. In most cases, the rifle will not fire. But if things are loose enough, and the conditions are just right, it may do just that. This would not be a true slam fire, but it could be just as exciting. This was a very, very, limited test. It did however, raise a few eyebrows.

We will now discuss jams, and pass on to you little gems of information that were passed on to us.

The M1 Rifle has an excellent record, and if kept clean, well lubricated, and fed good ammunition, it is soon enough found that most jams are self induced.

In some of the very early rifles that were produced as we entered WWII, some short cuts were made to the inside of the receiver that would later result in what became known as the seventh round jam. This was a case in which the rifle sometimes failed to feed the seventh round. Sometimes it even tossed the seventh round clear out of the rifle. An investigation was started, and it was soon found that if the first round loaded into the eight round clip was placed on the opposite side of the clip as it was filled, the jam went away.

Another jam that can occur is the bolt riding part way over the first round as the eight round clip hits bottom when loading the rifle. Hitting the operating rod handle (bolt handle) with the heel of the hand in the normal forward assist manner will promptly lock everything tight.

The reason this jam occurs on some rifles is that they have too much clearance between the magazine follower and its guide slots in the receiver. This in itself is not considered to be a serious problem as there is an easy way around it. Another way of course, is to have an armorer fit several new followers until one is found that fits tighter and does not have as much rear up-and-down free travel.

This jam problem first surfaced in the early days when the army was changing over from the 1903 Springfield. To load the magazine on the 1903, troops were taught to place their thumb at the rearmost part of the cartridge case, and as close to the stripper clip as possible. This ensured good smooth charging, for if the thumb was placed far forward on the case, the rounds would bind together and not leave the charging stripper clip. So, when the M1 was issued to the first troops, they of course placed their thumb well to the rear of the en bloc clip and shoved down hard.

The magazine follower on the M1 has its guiding slots 3/4 of the way to its front. The rear of it is totally unsupported and some can have over 1/8 of an inch of up-and-down free play. This can allow the clip to tip downwards at the rear far enough for the bolt to pass over the top round when the clip is pushed hard to the bottom of the magazine.

It did not take the Army long to realize that they had a problem with some rifles. So they began to teach that one's thumb must be placed even with the forward end of the clip. This cured most of the problem until the war broke out, and rifle production went into high gear. Then every place that they could loosen up the tolerances to speed up production was taken advantage of. Also, the manuals showing the correct thumb placement were rolling off the presses.

Big surprise! The first round jam began to surface again.

The Army, and also by this time, the Marines, took a good look at the M1 and how they were loading it, and now began to teach that the thumb must be placed farther forward on the cartridge case. If it was placed right at the case's shoulder on the top round of the clip, two things happened, both of which were good. The first was that the thumb was now exactly in line with the follower's guide slots and pushing straight down. The follower did not bind in its slots, and friction resistance was decreased. The second good thing was that placing the thumb far forward tipped the rear of the clip upwards and put the top round just where it should be when the clip hit bottom and the bolt was automatically released.

The absolute beauty of this solution was that it now no longer mattered whether the follower had tight, medium, or loose clearance tolerance in its guide slots; this method worked on all of them.

Thumb placed at cartridge's shoulder. A good push down from here will load the M1 correctly.

Thumb placed too far to rear of clip. This could cause a jam.

If you are looking for something to do on a dull afternoon and you have an M1 handy, try sticking a clip of empty or dummy rounds into it. Then try moving the thumb forward and back. It can be a real eye opener. Just watch out for the M1 thumb!

Sitting in the corner of the workshop is an M1 that has a stainless .30-06 barrel from one of our better makers fitted. This was turned from a solid blank many years ago. Never could get that poor thing to print under two minutes, and I always figured I had screwed it up somewhere. Also, every time it was removed for inspection, it had more indicator run out than the last time it was logged. Clearly we had some stress here.

Enter 300 Below of Decatur, IL (217) 423-3070 and their trio-lax stress relief program. The M1 barrel and a standard Colt .223 H-BAR barrel were shipped to them. When returned, they both had less run out than before. A skim cut was taken off of the M1 and it is now only out about .003.

Both rifles now print one MOA and sometimes less. There are a lot of things stress relieving cannot make up for. But this was not one of them. It is kind of a good feeling that the barrel was not chopped up and made into something else, as was being discussed. Thank you, 300 Below.

Not long ago, we were watching a program on TV about the evolution of firearms. All in all, it was quite good, and very informative. Right up until they got to the M1 Garand. Then the narrator solemnly informed everyone that when the M1 went empty, the ringing sound of its ejected clip gave away the empty rifle of the infantryman. He then went on to tell how wise infantrymen would carry extra empty clips which they would throw out to fool the enemy in Korea.

At this point, my big brother, who had fired an M1 in anger from one end of the Korean War to the other, came unglued. This is the same big brother who accidentally plugged our Grandma Rice's big radio while practicing quick draw with a big Colt double action black powder 44-40 when we were much younger. (See the *1995 Precision Shooting Annual* for that story.)

Big brother told the TV narrator that any stupid SOB who thought he could get close enough to the muzzle of an M1 to hear the clip after the roar of the eight rounds was sure welcome to come to his house trailer across the Ohio River from Wetzel County and test his theory. In a brilliant show of support, I sat quietly and watched our TV... praying silently that it would have a longer lifespan than Grandma's radio.

It is kind of hard to say how stories like this get started. But they do tend to live on among those who read more than they shoot. This is kind of like Gary Cooper licking his thumb, then rubbing it over his front sight to take the haze off of it, in the movie, "Sgt. York". Most any old shooter could tell you that what York did was wet his thumb, then press it into the dirt, then rub it across the sight, where the dried mud quickly dulled any shine. But then, a lot of things tend to get lost in the translation sometimes.

The search for accuracy never stops.

Update for the M1 Garand

Buying a Used M1

As I write this in 1999, a lot has changed and a lot has stayed the same. You can now buy an M1 at a lot of places. Most require an instant background check. A good place to start is the CMP (Civilian Marksmanship Program at Port Clinton, Camp Perry Ohio. Phone 888-267-0796 or fax (419) 635-2573). If you decide to buy from an

The author with his M1. No other words are needed.

individual, or at a gun show, always remember to "let the buyer beware".

At a recent show, I looked over a table labeled "US Government Match Garands". All were standard issue GI models. When questioned about this, the owner stated that "All were made for the US Government, and all had been fired in a match at one time or another." So, if you are looking for an accurized model, or a National Match Model, what are some of the things to look for?

First, most, but not all, match models will usually have a hooded rear sight. If the rifle has been accurized, the front sight and gas cylinder should be a tight fit on the barrel, and you should not be able to move it with your hands. A standard GI model will be fitted loose so it will interchange with all of the rest out there.

If the rifle passes these quick tests, it is time to go on to some others that are much harder to fake. The trigger pull can be weighed, or checked with an accurate spring gage. It should not be under $4^1/_2$ pounds, or over 5 pounds. Most GI "issue" rifles are over this.

The correct fitting of the operating rod (bolt handle) to a match Garand can make or break it accuracy-wise, and is a job that takes time and know-how. To check this, close the bolt, hold the rifle by its pistol grip and rest the bayonet lug on a waist-high table. Grip the front handguard with your other hand, reach under it with one finger and push up on the op rod just behind the gas cylinder. It should raise freely, just a tiny bit, until it stops at the top of the gas cylinder, and when released, it should fall back freely. If it does not do this, it is probably binding somewhere and applying stress to the barrel. The chances of it shooting well are not good.

A truly accurized rifle will be glass or epoxy bedded. Some sellers may object to taking the rifle apart to show you this, and that is OK, for it should not be pulled apart without good reason.

However, they should not object to pulling the trigger group out and trying to move the rear of the action up and down slightly in the stock with the tips of your forefinger and thumb. For this is the accepted method armorers use to see if the bedding needs to be tightened up with a fresh application of compound. With this test, a tightly bedded action cannot be moved with just the finger tips. A loose one can, and a standard issue GI will usually fall right out of the stock.

Even if the rifle passes all of these tests, it could have had one heck of a lot of rounds through it over the years. So by far the best test is to shoot it. If the seller wants top dollar for his rifle, he should be proud of it, and eager to show you what it can do.

A good accurized M1 should group 5 to 10 shots in a 2-inch circle with match ammo at 100 yards off of a bench rest. If it does better than this, kiss him on both cheeks, give him the money, grab the rifle and run like hell. There are some 1 MOA M1s out there, but they are rare.

A long time ago, some armorers would stamp a lot of the parts they used "N.M." for the words "National Match". It did not take some sharpies long to learn that buying stamps and stamping all of their used parts could get them a few extra bucks. Over the years, so many rifles were accurized that the practice has pretty much passed on. If, however, you run across one so marked, and it passes all of the other tests too, you probably have a genuine one. But the final proof is in the grouping. If it won't do this, it could need a new barrel.

Reloading

All of the information on reloading covered in this book is still valid. There are some new bullets coming on the market each year. Most are held to much closer tolerances than ever before, and this is great. The best loads to use in the M1 are still those that come close to the US Govt. N.M. 173 grain round.

IMR 4895 and H-4895 are still excellent powders. And Accurate Arms #2520 is a good ball powder. Generally speaking, the M1 usually does its best with powders like this. If you have the slightest doubt as to a powder being too fast, or too slow for an M1, call the manufacturer, whose name should be on the container. And if their phone number is not there, call your local library and ask for the Business, Labor, and Government division. They will probably get you their number. You can also call the NRA and most of the bullet making companies. Sierra is very good on this.

Bullets heavier than 180 grains usually do not do as well as a good match grade 175 grain. If you try to get maximum velocity out of the heavy ones, their recoil beats both you and the rifle's bedding, and you do not get a great deal in return.

There is still a lot of corrosive primed ammunition out there and I am very much against shooting it in anything but a junk rifle. This is kind of like driving a new car over salted roads in the winter. You can clean it, and clean it, but you will never get it all. And eventually you will lose. The same thing happens with an M1 or any other rifle. The salt gets ironed under the jacket fouling in the barrel, and the M1's gas system sprays it out the back of the gas cylinder.

There are those who will tell you that the M1 was designed for this, and it is OK to shoot it. The fact is Mr. Garand was forced to design a rifle that could live with corrosive priming because that was the most reliable primer that they could get for war at that time. The warehouses were full of it, and this was not about to change for a long time. The difference in cost of corrosive versus non-corrosive ammo is seldom worth it in the long run.

The main reason the gas cylinder (and front sight) was fitted so loose to the barrel was that it had to fit all the other barrels out there, and in combat in WWII it could rust on so tight that it had to be literally pounded off so you could clean the rifle in order to have it function reliably.

Cleaning and Lubricating

If the rifle was fired with corrosive ammunition, it should be cleaned with GI bore cleaner. You can get by with hot soapy water followed by a good cleaning with a commercial cleaner like Hoppe's. But you better check the rifle from top to bottom for the next few days if you want to avoid trouble.

If you fire normal ammunition, clean the rifle in the normal way. Just read the directions on the bottle. Most armorers clean the rifle while it is held in a cradle, with its barrel horizontal with its sights up. This way, some cleaning solvent flows through the gas port to soften any carbon fouling there. They then remove the gas cylinder lock screw and clean the gas cylinder the same way they do the barrel. They do not remove the gas cylinder to clean it. Today, most use Hoppe's or Shooter's Choice. It is kind of a toss up as to who you talk to.

An M1 Garand must have its bolt's locking lugs and their camming surfaces lubricated or sooner or later, they will start to wear and scuff. Any good lubricating oil is better than nothing, but the military always leaned towards the white grease, Lubriplate. This is an excellent choice, for as soon as it gets dirty, you can certainly see it.

Still, I have talked to others who use a moly grease which is usually dark to start with, and they just change it often, at about 200 to 500 shots.

The way to change the grease is to just wipe the old off with a clean rag. Do not degrease with a solvent, for if you miss a place you will do more harm than having the old lube there.

Along with the locking lugs, also do the tracks in the receiver that they ride in, the cam in the op rod that opens the bolt, the nose of the hammer where it rides against the bottom of the bolt, and the shiny ends at the rear of the bolt. Also put some on the bottom of the barrel if the op rod is rubbing here.

The Department of the Army used to distribute a manual for the care and cleaning of the M1 called FM 23-5 US rifle, caliber .30, M1. This book started out pretty big, but as time went by it got smaller. The small, soldier's version is fine for most people. CMP, at Camp Perry, may still have some. I see them at gun shows all of the time, so gun show book dealers should have some. If not, someone will surely be reprinting them. It is a good book to have.

Some Passing Thoughts

As I write this I am well aware that I am fast approaching my 69th birthday. Looking back over my life, there is little that I would change.

There have been men in our family tree involved with rifles going back to Lancaster, PA, to Germany, and on back to Switzerland. Recently we have been retracing the footsteps of a great grandfather who was at Antietam with the Pennsylvania Cavalry.

My earliest recollection of Camp Perry is sitting in a Model "B" Ford with my mother while Pop competed. My big brother was allowed to stand by Pop's extra equipment behind the assembly line. In later years I was allowed to tag along behind Pop, Uncle Roy, and brother Dale. I was entrusted with the extra bandoleers of ammunition that should not be needed. I was probably about 7 years of age back then.

After growing up, I became an Aviation Machinist and got involved with sports cars. I was able to road race motorcycles on the Canadian Grand Prix circuits when the English single cylinders were king.

When I was 29, I met a young hillbilly gal who had just turned 20. We were married, and out of kindness, our best friends gave it about a

Mary with her Mauser 6mm Rem. Sometimes she would say things like, "Honey, you just ain't got what it was."

year, most thought less. 37½ years later it was still getting better every day.

Mary Alice and I were not just a husband and wife. We were always best friends. We raised two daughters who are now happily married with children of their own. She was the best pit crew a racer could ever want.

Somewhere around age thirty, she began to go with my Dad and me on woodchuck hunts, while my mother camped in an old farm house with the kids. This led to Pop building her a 6mm Remington on a Mauser 98 action. This in turn led to yearly deer hunts in Wyoming where one of Pop's friends lived. After going through the DCM school at Camp Perry, she applied for an M1, and to everyone's amazement, they sent her a brand new one. These two rifles were to become her prized possessions and we never went many places that one or both did not go along.

In 1995 Mary was diagnosed with Pancreatic Cancer and in 39 days she was gone. I would like to dedicate part of this book to her, as most of the pictures of hands holding parts, etc. were hers.

She was always there, looking over my shoulder, and when laughing or excited, she could totally destroy the English language. At times like this, if you told her something she did not agree with, she would come back with, "Honey, you just ain't got what it was."

So what I am trying to tell you here, is that if you have someone you care about, and they are not interested in the things you are, do not push the issue. Just answer their questions, and you may be surprised some day.

A while back, I was working with a woman on some magazine articles. She was excellent to work with, and had two Master Degrees. She had spent thirty-some years as a Nun. After leaving the Order, she taught school for the remaining years, until her retirement.

She was always looking for some new challenge, and so began doing research on the training a female officer goes through to become a police officer. One day, out of the clear blue, she looked from our work and asked me to teach her to shoot a rifle and a pistol. The next day found us at the range, where she took tons of notes, as we went from a .22 rifle to a .22 pistol, and on up to what a female officer would be expected to qualify with. I can assure you that you have not lived until you have met a group of friends for supper, and when asked what you have been doing with yourself, you casually dismiss the question with, "Oh, just teaching a nun to shoot a Colt .45 automatic."

The search for accuracy never stops.

The Iron Sight Sniper

(*Tactical Shooter,* December 1998)

Over the years you hear many stories about WWII. Some are so full of inaccuracies that they can be discounted right from the start. But every once in a while you hear one that rings with truth. This is one of them.

Many years ago a gentleman I had learned much from while working in an aviation machine shop ended up in a hospital from old age and other complications. While visiting him on a rainy afternoon, Herman told me about his M1 rifle, and this story:

Herm's Story

I grew up on a dairy farm in Iowa. My father was the secretary of the local shooting club, and I learned to shoot the .03 Springfield all the way back to 500 yards, as that was as far as we had room for on one of the farms.

When the war started, I joined up right away. When the Army found out that I could keep trucks running, and was a good mechanic from all of the years working on farm machinery in the fields, they kept me in the US.

After the invasion of Europe, our unit was sent there. At first, I

The M1 rifle. Because of its excellent sights, it gave good long range accuracy.

wanted to get into combat, but after seeing so many badly wounded returning from the front for so long, I decided to just do what I was told.

Then the Battle of the Bulge started, and we were moved closer to the front lines. It was a terrible winter, and we did most of our work outside, wherever the disabled vehicles were. Sometimes we could spread a pup tent over the part we were working on and with a kerosene lantern we could crawl in under there and have enough light to work with, and keep our frozen and cut hands partly warm, too.

By the time the Germans were being held for good, the casualty rate was so high on our side, that they sent most of my outfit up to the front to be replacement riflemen. I had been handling M1 rifles for quite a bit, but had not fired one in a long time. I really loved that rifle, and the sights on it were so much better than the old .03 I had grown up with. It was hard to load when you were frozen stiff, and any time you got down to a few rounds left in the rifles, you just shucked the rest. You always kept a full clip in it, unless you were on guard duty, then you kept 7 in the clip, and the bolt closed on an empty chamber.

I was put in with a group of replacements, and we were sent in to a part of the lines that was held by Paratroopers, and mixed in with them, wherever we could be used.

I was assigned to a squad that was all troopers, and they pretty much kept to themselves and I was the outsider. Part of this was because they had lost so many of their men, that they tried not to let themselves get too close to you, then when (not if) you got it, it would not bother them so bad. Along with this, they had been fighting ever since they jumped into Normandy, and did not think much of people who had spent all of this time working on trucks.

Sergeant Harris was pretty good, and tried to do the best he could for his men. I was put in a foxhole with a long, lanky trooper from Kentucky who was called "Kaintuck" by all of his friends, and after a few nights and days together, we began to get along OK. But if we were sent back to the farmhouse cellar to get some food and sleep, he went with his friends.

One morning, just as it was getting light, we were in our hole, and Tuck was dozing. I caught some movement in the trees to our left front, and saw the whole German Army coming. I yelled "Tuck!" as loud as I could, and it came out almost like a squeak. Then I started firing, and got two of them from the side before they knew where it was coming from.

It was hot and heavy, and I could see Tuck out of the corner of my eye. He was firing as fast as he could, and I never looked at him again, until I was digging into my second bandoleer for more loaded clips. We were both scared to hell and gone.

Our reserves rushed in, and we finally pushed them out of the woods and off of our hill. As things eased off, it dawned on me that this was the first time I had ever fired this rifle, and that I had no zeros for it. (I guess your mind works kind of funny like that.) So I told Tuck to watch the snow on the side of a tree about 100 yards down the hill in front of us, 'cause I was going to check my sight's zeros on it.

Without waiting for his reply, I took careful aim and let a round go. The windage was spot on, but the elevation was low. So while Tuck was mumbling about being stuck in some front line hole with a nut, I brought the elevation up 7 clicks from the bottom. I still remember that after all these years. After the third shot to confirm, Sergeant Harris yelled over and wanted to know what the hell we were shooting at. Tuck yelled back that I was zeroing my rifle, and it got so quiet, that you could hear Harris mumbling. And way off to the right I could hear someone laughing.

We were issued more ammo, and got all set for another attack. But the morning wore on, and it never came. Later that night when we got to go back to the cellar to eat, Kaintuck and the guys asked me to sit on their part of the dirt floor with them. It felt so good I started to cry, but no one even let on.

A few days later at the hand pump at the well in back of the farmhouse wall that was still standing, a trooper asked me if I was the crazy civilian truck driver that zeroed his M1 in the middle of the battle.

They hit us again two mornings later, came right up through the very same spot. But we were ready for them, and Sergeant Harris had moved two air-cooled .30's into the area. With the help of those two machine guns, we beat them back again. Again, they did not counter-attack, which was rare, and looking back now, I think they just did not have the strength.

A few days after this, Sergeant Harris called our squad together and told us that the Lieutenant was putting a patrol together, and that we were it. We met Lieutenant Morris around behind the back of a hill and he told us that we were going to probe well off to our right to see if the enemy was growing stronger there. And we were not to fire unless we had to, or were told to. We moved through the hills and woods for some time, and in places the snow was over your boots. Then we halted well behind one of the bigger hills, and after a brief rest, Sergeant Harris spread the men out in a defensive position. I was left standing near the Lieutenant.

When Harris returned, the Lieutenant motioned for us to follow, and set out making his way to the left side of the top of the hill. We crawled the last part of the way, and with the M1 cradled in my arms, I was constantly shoving snow in the cuffs of my overcoat. This ran down my arms and was cold as hell.

There was an old blown-down tree just at the ridge, and we crawled into the branches. It was a pretty good hiding place, as all of the branches broke up our outlines.

Well out in front, down in a kind of hollow, was another farmhouse. It was in pretty good shape, compared to most, and the Lieutenant, using about the biggest pair of binoculars I had ever seen, said he could see men in positions about the outer buildings.

He passed the glasses to Harris, and they were discussing things when we heard motors and saw two motorcycles with a civilian sedan behind it pull into the yard. The driver got out, and opened the rear door for the officer and he got out and began to move about, right out in the open, speaking to men who were under cover. It was a high ranking officer come up to inspire his men before a big attack.

Lieutenant Morris took the glasses back and whispered to me, "Do you think you can hit that bastard from here?" None of us could believe what we had stumbled on to.

Without speaking, I slid the M1 through the branches until it rested on my palm, which lay on top of the main trunk. I figured the wind for almost nothing, and cranked the elevation up to 500. Lieutenant Morris could see them pretty good through the glasses, but I could not tell which was which, and to this day I am not sure how far it really was, 'cause we sure didn't stick around for the target to be scored after the shot.

I told the Lieutenant I could not tell who was who, even when he said, "Don't shoot any of the ones wearing helmets". Along with this, as the man kept moving, he put branches and limbs between us, and I could not get a clear picture through the sights.

Then other men, the Lieutenant said he thought they were also officers, came out of the woods and he said he thought they were trying to get him to take some cover, as they greeted him, but he just dismissed them with a wave of this hand, and at that same time stepped into a clear spot between the branches for me. I was sure I had the right one, as no one else seemed to be allowed too close to him. I was just putting the front sight on him, and taking up the slack in the trigger, when the Lieutenant whispered, "That's him, right in line with the edge of the white barn wall". I tripped the trigger, and the big M1 barked with authority, and slammed into my shoulder.

"You got him!" the Lieutenant said excitedly, almost out loud. "You got him." Then, a few seconds later, "Come one, we got to get out of here... it's a hornets nest over there."

We slid rapidly back down the hill until we were out of their sight. Then the Lieutenant and Sergeant Harris stood up and began running down the hill. I was trying to keep up, but was lagging way behind, as I was wearing big 4-buckle rubber boots. Behind us, we could hear big mortar rounds hitting the other side of the hill to the right, as they searched for us.

The Sergeant got the men started back the way we had come on a dead run as bigger stuff joined in and swept our way. Lieutenant Morris, seeing I was still lagging behind, came running back and grabbed me by the front of my cartridge belt and pulled me along. We literally made giant steps as we ran through the snow and woods.

Back in our regular lines that night, we could feel the impact as the Germans pounded those woods and hills. The word quickly spread that I had shot a high-ranking officer. By the next day the story going around was that I had plugged their commanding General from over a mile away with an issue M1.

Things settled down almost to normal a few days later and Kaintuck and I were sent to the rear to help carry up a load of rations. He loaded down my pack with glee, and went over to get some more for his, stopping to talk with some other troopers that had just brought a prisoner out of interrogation. As we made our way back with the others he kept smiling like a cat in front of a bowl of warm milk, and kept telling me what a pleasure it was to know me. I figured something was up, but could not guess what.

Back in the cellar we dumped the rations in a pile, and Sergeant

Harris began to distribute them. As he did so, he casually asked Kaintuck if there was any news from the rear.

Seated with his back against the cellar wall, Kaintuck savored this question for a full moment or so. Then he nodded towards the German lines and said with a sly smile "They's pretty mad over there...." He stopped for a moment to swallow a mouthful of canned meat and potatoes. "Yep, they's pretty mad over there... one of their patrols bumped into one of ours last night and we took 'em all prisoners. One of 'em said they was all up in arms... 'cause... someone... shot... their... Lutheran... Pastor. Yes sir, he come up there to see his people... and some damn Yankee done shot him right in the ass. That medic said the bullet done went through both cheeks just as he was turning to say hello to their commanding officer."

In the half light of the cellar, I could see all of them looking at me. It was quiet for almost a full two seconds. Then the place turned into a mad house. One threw his helmet against the far wall, and everyone was laughing so hard the tears were streaming down their dirty faces. Next to me, Kaintuck slammed my knee again and again as I sat with my mouth open.

Finally, as the laughter settled down, Sergeant Harris stood up. "Don't nobody say anything to the Lieutenant. This one is all mine." He turned and went up the small steps, and I remained the center of attraction.

Kaintuck leaned back against the wall and sighed. "Damned if you ain't the damnedest iron sight sniper I ever knew."

The nickname stuck in our small unit. It felt a lot better than "Civilian Truck Driver", in this group of fighting men.

The M1 and the Resurfacing Nambu

(*Tactical Shooter,* August 1999)

This story was told to me at a motorcycle club awards banquet some time back. I had worked with George at some motorcycle events, and considered him a very good man. So I will pass it on, as it was quite interesting.

George's Story

The fighting on Guadalcanal had been going on for some time before the infantry unit I was with arrived during WWII. The Marines had absorbed many of the Japanese during the early part, and our army had been pounding them for some time, while we were being used to unload and distribute supplies as they came in from the ships. Finally the island was considered pretty much secured and they began to send some of our unit out to find and mop up scattered small pockets. I had not turned twenty yet.

By this time though, we had all been in small actions, and had been on the island long enough to have various cases of malaria and dysentery, so it was not like we were green troops. Most of us had come to realize that the M1 was one hell of a good fighting rifle, and we relied on it as much as we did on each other.

Our sergeant called us together after supper, and we all went over and filled our cartridge belts, plus we each drew two bandoleers. The next morning at assembly, the lieutenant took over the squad and we set out for the hills. There were good trails for awhile, but they finally got fewer and fewer, and the first night was spent by a small stream.

You never really sleep in the jungle, so by midday we were dragging up the stream, pretty well spread out, which was OK, and I had drawn the tail end position. As such, it was my job to keep an eye on the trail behind us. So I had to keep turning every couple of steps to watch. If it was anywhere near level and not too rocky, I could sometimes walk backwards for a few steps, with the M1 always cradled in my arms, or held ready in my hands. As it turned out, this and about a foot of water saved my life.

We had reached a place where the land was pretty level and the stream bed was reasonably wide, but the jungle on both sides was still too thick to get through, when the point man found kind of a pig trail coming in from the left, and it had footprints. So he and the sergeant went down it for a ways.

The lieutenant and the sergeant held a huddle, and as it looked like whoever had used the trail had gone upstream and there did not look to be more than about four of them, the decision was made to follow, but very carefully. We were all on full alert, and everyone expected an ambush at any second. The guy closest to me was my buddy, Bob Homan, and we were both scared to death.

The squad had reached a bend in the stream and most of them were already around it. We were maybe one hundred yards up from the pig trail, and I was walking backwards for my couple of steps when two

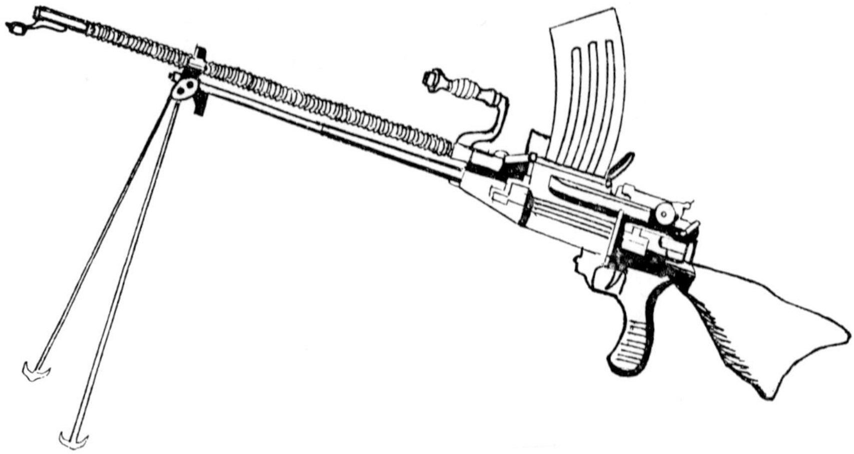

The Model 1936 6.5mm Nambu.

Japanese soldiers ran out of the trail and flopped down in the gravel by the stream. They had a Nambu light machine gun. There wasn't time to even think, and Bob said later that I yelled his name as loud as I could, but I sure don't remember it.

I hit a prone position right in the stream, and the water was almost over my mouth. The first round went over them, I am sure, but I got a good sight picture on the next three. They fired a good long burst in which their rounds went over me and blew water spouts all around me. I got off the rest of the clip. Then they stopped firing and one of them helped the other back into the jungle. They left the Nambu.

Bob had been firing like crazy all this time, and got off one and a half clips. Part of the squad had heard my yell, plus all the shooting, and some of the Nambu 6.5 rounds had been tearing into the bush where the stream turned. We took cover around the stream's bend, and the lieutenant and sergeant held another huddle. The lieutenant felt sure that all of the shooting would bring someone from upstream.

Bob and I were crammed in under the stream bank's edge, and by raising just enough for our eyes, we could see the Nambu, still sitting on its bipod in the gravel by the stream. Everything had happened so damn fast, probably way less than sixty seconds. Bob whispered that he had got off most of his rounds after they had jumped up and ran for the jungle's edge, and could not believe that they gave me all of their fire, when he was completely exposed, prone in the gravel. Maybe it was because I was already shooting at them.

The lieutenant made a quick decision, and we moved upstream to a slightly better place and set up a really good ambush. We stayed there for quite a while and when nothing developed, we moved upstream again. The next day we probed up to where the stream became too narrow and small to go on, so we started back.

When we got to where the fight was, and eased around the bend, the Nambu was still sitting there. Bob and I were picked to ease up on it and check things out. He found some blood on the pig trail, and I found that one of my bullets had hit the Nambu right where the magazine goes in from the top. This bent things so bad that you could not get the magazine out, and the bolt would not close.

Sarge told us to pitch it, so we tossed it into the jungle on the other side of the creek. Scratch one model 1936 Nambu.

The Resurfacing Nambu

After the war, I came home, got a job and got married. The second company I worked for made mining equipment.

We were at an industrial show and at dinner I was seated next to a salesman who handled our stuff in Australia. He had been doing business with an engineer whose company was working on Guadalcanal. When I mentioned that I had been there during the war, and would like to go back someday, he gave me the guy's address and number.

Then he said, "He knows a real good guide there. If you ever get

The M1 rifle. For many soldiers, it meant coming home... or not.

there, be sure and ask to see his old Jap Nambu machine gun. It's all bent to hell where it was hit by a bullet. You can't get the magazine out of it, and it won't fire anyway. He and his cousin found it up in the hills after the war. It took them a day and a half to lug it all the way back down a little stream."

Small world. I sometimes think back about that M1 that saved my life. You know, by the time we were pulled off the island, the Marines had exchanged enough of their 1903 Springfields to where it was safe to set your M1 down for a while. Hell, you didn't even have to sleep with it cradled in your arms anymore.